The Indus Basin of Pakistan

DIRECTIONS IN DEVELOPMENT
Countries and Regions

The Indus Basin of Pakistan

The Impacts of Climate Risks on Water and Agriculture

Winston Yu, Yi-Chen Yang, Andre Savitsky, Donald Alford, Casey Brown, James Wescoat, Dario Debowicz, and Sherman Robinson

THE WORLD BANK
Washington, D.C.

Library of Congress Cataloging-in-Publication Data
Yu, Winston H.
The Indus Basin of Pakistan : the impacts of climate risks on water and agriculture / Winston Yu, Yi-Chen Yang, Andre Savitsky, Donald Alford, Casey Brown, James Wescoat, Dario Debowicz, and Sherman Robinson.
     p. cm.
   Includes bibliographical references.
ISBN 978-0-8213-9874-6 — ISBN 978-0-8213-9875-3 (electronic)
   1. Climatic changes—Pakistan. 2. Crops and climate—Pakistan 3. Water resources development—Environmental aspects—Pakistan. 4. Indus River Watershed—Economic conditions. 5. Indus River Watershed—Environmental conditions. I. World Bank. II. Title.
   QC903.2.P18Y82 2013
   551.48095491—dc23   2013008111

Contents

Figures

Maps

Tables

Acknowledgments

This volume was prepared by a team led by Winston Yu (World Bank). Yi-Chen Yang (University of Massachusetts, Amherst) and Casey Brown (University of Massachusetts, Amherst) generated model scenarios for the Indus Basin Model Revised (IBMR); Andre Savitsky (consultant) updated the IBMR and improved upon the original GAMS (General Algebraic Modeling System) code. Donald Alford (consultant) provided an analysis of the snowmelt and ice melt hydrology in the upper reaches of the Indus Basin; James Wescoat (Massachusetts Institute of Technology) provided an overview of the institutional and policy issues on water and agriculture in Pakistan; and Dario Debowicz (International Food Policy Research Institute) and Sherman Robinson (International Food Policy Research Institute) developed a computable general equilibrium model for the economy of Pakistan. Masood Ahmad (World Bank) and Paul Dorosh (International Food Policy Research Institute) provided invaluable technical assistance and guidance throughout this study. Their generous contributions have lent clarity and focus to this work. The authors are also grateful for the support of the World Bank management team and the Pakistan country office, including Rachid Benmessaoud (Country Director), Simeon Ehui, John Henry Stein, and Gajan Pathmanathan. The authors extend their sincere thanks and appreciation to John Prakash, Venkatakrishnan Ramachandran, and Shabir Ahmad for their administrative support. Editorial support of Valerie Ziobro and Natalie Giannelli is gratefully acknowledged.

The authors benefited enormously from the many technical discussions with colleagues and officers of the Water and Power Development Authority (WAPDA), Government of Pakistan: Irshad Ahmad, Syed Raghib Abbas Shah, Allah Bakhsh Sufi, Daniyal Hashmi, Abdul Rehman, Muhammad Siddique, Iabal Masood Siddiqui, Mushtaq Ahmad, Rao Shafiq-ur-Rehman, and Shafeeq ur Rehman. Useful guidance during the early stages of the study was provided by Shams Ul Mulk (former WAPDA Chairman), Ishfaq Ahmad (Planning Commission), Zafar Altaf (Pakistan Agricultural Research Council), and scientists from the Global Change Impact Studies Centre (GCISC). These discussions helped to shape the approach developed in this study.

The authors also acknowledge constructive comments and suggestions from the following reviewers: Julia Bucknall, Nagaraja Harshadeep, Daryl Fields,

Marinos Tsigas, Luis Andres, Usman Qamar, Rita Cestti, Aziz Bouzaher, Madhur Gautum, Mahwash Wasiq, and Tahira Syed. Their contributions have enhanced the quality of this document, which is intended to support Pakistan in its efforts to better manage climate risks in the water and agriculture sectors.

Generous support was provided by the World Bank, the Bank Netherlands Water Partnership Program, the Bank–Dutch Trust Fund for Pakistan, and the South Asia Water Initiative.

About the Authors

Winston H. Yu is a senior water resources specialist at the World Bank. He has extensive experience in technical and institutional problems in the water sector and has carried out a number of research and investment projects in developing countries. His special interests include river basin management tools, hydrologic modeling, flood forecasting and management, groundwater hydrogeology, international rivers, and climate change. Prior to joining the World Bank he was a researcher at the Stockholm Environment Institute and served as a Science and Technology Officer at the U.S. Department of State. He is also an Adjunct Professor at the School of Advanced International Studies at Johns Hopkins University where he is associated with the Johns Hopkins Global Water Program. He received a PhD from Harvard University.

Yi-Chen E. Yang is a postdoctoral research scientist at University of Massachusetts Amherst, and recently received a doctoral degree in Civil and Environmental Engineering from the University of Illinois. His research interests are water resources management, water policy development, climate change impact assessment, coupled natural and human complex systems, agent-based modeling, decentralized optimization, hydroecology, and geographic information science. He has a BS in geography and an MS in bio-environmental system engineering from the National Taiwan University.

Andre Savitsky is an independent consultant. He has published more than 50 articles on hydrology, hydrodynamics, and water resources management and continues research in these areas, specifically in the wide usage of mathematical modeling and programs for water accounting and management internationally. He studied hydrology at the State University of Uzbekistan and mathematics at the Tashkent Pedagogical Institute, and received a PhD in hydrology and hydraulics.

Donald Alford is an independent consultant. He has studied mountain glacier and hydrometeorological environments ranging from northern Greenland to Southeast Asian subtropics. His Himalayan studies began as a visiting scientist at the East-West Center in Honolulu and at the International Centre for Integrated Mountain Development in Kathmandu. Recent studies have involved the relationship of climate change, glaciers, and water resources in the Pamir and

Hindu Kush–Himalaya regions. He studied earth sciences at Montana State University and the Institute of Hydrology and Glaciology at the University of Zurich and obtained a PhD with a dissertation on cirque glaciers from the Institute of Arctic and Alpine Research, University of Colorado.

Casey Brown is an assistant professor in the Department of Civil and Environmental Engineering at the University of Massachusetts at Amherst and adjunct associate research scientist at Columbia University in New York. He has worked extensively on projects around the world in climate and water resources. He has received awards, including the Presidential Early Career Award for Science and Engineering, the National Science Foundation CAREER Award, and the Huber Research Prize from the American Society of Civil Engineers. He has a PhD in environmental engineering science from Harvard University and led the water team at the International Research Institute (IRI) for Climate and Society at Columbia University.

James L. Wescoat is the Aga Khan Professor in the School of Architecture and Planning at the Massachusetts Institute of Technology. His research has concentrated on water systems in South Asia and the United States at the river basin scales. He has contributed to studies of climate, water, agriculture, and energy in the Indus Basin, and to historical research on waterworks of the Mughal period in India and Pakistan. He served on the U.S. National Research Council's Committee on Himalayan Glaciers, Water Resources, and Water Security. Wescoat received his PhD in geography from the University of Chicago.

Dario Debowicz is a postdoctoral researcher at the International Food Policy Research Institute in Washington, DC. He has lectured on macroeconomics at the University of Buenos Aires and other universities in Argentina and on development economics at the Institute of Development Studies. Debowicz conducts macro-micro countrywide modeling and has worked with computable general equilibrium (CGE) models, dynamic stochastic general equilibrium models, and behavioral micro-simulation models. His research interests include development macroeconomics, monetary policy and international finance, labor economics, the macroeconomics of poverty reduction, and the partial and general equilibrium effects of conditional and unconditional cash transfers. He received his PhD in economics from the University of Sussex.

Sherman Robinson is a senior research fellow at the International Food Policy Research Institute in Washington, DC, and professor of economics (emeritus) at the University of Sussex, U.K. He is a leading expert on CGE simulation models, and his research interests include international trade, economic growth, climate change adaptation, linked water-economy models, macroeconomic policy, income distribution, and maximum-entropy econometrics applied to estimation problems in developing countries. He has published widely in international trade, growth strategies, regional integration, income distribution, empirical modeling methodologies, and agricultural economics.

Abbreviations

ACZ	agro-climatic zone
AF	acre-foot
AR4	Intergovernmental Panel on Climate Change Fourth Assessment Report
ASL	above sea level
ASP	Agricultural Statistics of Pakistan
BKWH	billion kilowatt-hours
BRW	Balochistan Rice Wheat
CANEFF	canal and watercourse efficiency improvements
CCA	culturable command area
CDF	cumulative distribution function
CGE	computable general equilibrium
CPI	consumer price index
CPS	consumer and producer surplus
CRU	Climatic Research Unit
CV	coefficient of variation
CYIELD	crop yield
DEM	digital elevation model
DIVACRD	1991 Provincial Water Allocation Accord
FAOSTAT	Food and Agricultural Organization of the United Nations Statistical Database
GAMS	General Algebraic Modeling System
GCISC	Global Change Impact Study Centre
GCM	general circulation model
GDP	gross domestic product
GM	genetically modified
GRDC	Global Runoff Data Centre
HKK	Hindu-Kush-Karakoram
IBDP	Indus Basin Development Programme
IBIS	Indus Basin Irrigation System

IBM	Indus Basin Model
IBMR	Indus Basin Model Revised
IBMY	Indus Basin Multi-Year
IPCC	Intergovernmental Panel on Climate Change
IRSA	Indus River System Authority
IWT	Indus Waters Treaty
JJAS	June, July, August, and September
MAF	million acre-feet
MASL	meters above sea level
MTDF	mid-term development framework
NEWDAM	construction of new reservoirs
NSIDC	National Snow and Ice Data Center
NWFP	North-West Frontier Province
NWKS	North West Kabul Swat
NWMW	North West Mixed Wheat
ONDJ	October, November, December, and January
PCWE	Punjab Cotton Wheat East
PCWW	Punjab Cotton Wheat West
PMD	Pakistan Meteorological Department
PMW	Punjab Mixed Wheat
PRs	Pakistani rupees
PRW	Punjab Rice Wheat
PSW	Punjab Sugarcane Wheat
ROW	rest of the world
SAM	social accounting matrix
SCARP	Salinity Control and Reclamation Projects
SCWN	Sindh Cotton Wheat North
SCWS	Sindh Cotton Wheat South
SRTM	Shuttle Radar Topography Mission
SRWN	Sindh Rice Wheat North
SRWS	Sindh Rice Wheat South
SWAT	Soil and Water Assessment Tool
SWE	snow-water equivalent
UIB	Upper Indus Basin
USEPA	United States Environmental Protection Agency
WAPDA	Water and Power Development Authority
WSIPS	Water Sector Investment Planning Study

Executive Summary

This study, *The Indus Basin of Pakistan: The Impacts of Climate Risks on Water and Agriculture*, was undertaken at a pivotal time in the region. The weak summer monsoon in 2009 created drought conditions throughout the country. This followed an already tenuous situation for many rural households faced with high fuel and fertilizer costs and the impacts of rising global food prices. Then catastrophic monsoon flooding in 2010 affected over 20 million people, devastating their housing, infrastructure, and crops. Damages from this single flood event were estimated at US$10 billion (ADB and World Bank 2010), half of which were losses in the agriculture sector. Notwithstanding the debate as to whether these observed extremes are evidence of climate change, an investigation is needed regarding the extent to which the country is resilient to these shocks. It is thus timely, if not critical, to focus on climate risks for water, agriculture, and food security in the Indus Basin of Pakistan.

Pakistan relies on the largest contiguous irrigation system in the world. Known as the Indus Basin Irrigation System (IBIS) for its basic food security and water supply for all sectors of the economy, it supports the basin comprising the Indus River main stem and its major tributaries—the Kabul, Jhelum, Chenab, Ravi, and Sutlej rivers. IBIS has 3 major multipurpose storage reservoirs, 19 barrages, 12 inter-river link canals, 45 major irrigation canal commands (covering over 18 million hectares), and more than 120,000 watercourses delivering water to farms and other productive uses. These canals operate in tandem with a vast and growing process of groundwater extraction from private tubewells.

IBIS is the backbone of the country's agricultural economy. The agriculture sector it supports plays a critical role in the national economy and livelihoods of rural communities. Agriculture contributes some 22 percent to Pakistan's gross domestic product (GDP), down from 27 percent in 1989 and 46 percent in 1960, due primarily to more rapid growth in the services sector. Forty-five percent of the labor force is employed in the agriculture sector. The value of agricultural production continues to grow at an average annual rate of approximately 3 percent. However, the inter-annual variability of agricultural value-added to GDP is high, which demonstrates existing vulnerabilities to climate risks.

Irrigated land supplies more than 90 percent of agricultural production. Agriculture in most areas of the basin is not possible without irrigation because Pakistan's climate is arid to semi-arid with low and variable rainfall. Only 28–35 percent of the total land area is arable, and that proportion has not increased significantly in recent decades. However, the irrigated portion of arable land has grown over the past decade—from about 65 percent in 2001 to almost 75 percent in 2009—contributing to increased agricultural production and yields.

The rivers of the Indus Basin have glaciated headwaters and snowfields that, along with monsoon runoff and groundwater aquifers, are the major sources of water for Pakistan. Currently, about 50–80 percent of the total average river flows in the Indus system are fed by snow and glacier melt in the Hindu-Kush-Karakoram (HKK) part of the Greater Himalayas, with the remainder coming from monsoon rain on the plains. Variability in the distribution and timing of snowfall and changes in snow and ice melt may be amplified by climate change, which has implications for managing basin water resources.

Managing groundwater resources continues to be a major challenge in the Indus Basin. Waterlogging and salinity have been major concerns over the past century, following the expansion of canal irrigation. Groundwater levels and water quality vary across the plains during the irrigation and monsoon seasons. When tubewells tap into brackish groundwater, they accelerate the secondary salinization of irrigated soils, which injures crops and reduces yields.

Food self-sufficiency is an escalating concern in Pakistan. Although agricultural production and yields continue to grow, the population growth rate also remains high, at an annual 2.2 percent. Per capita food supply varies from year to year and falls below the global average of 2,797 kilocalories (kcal) per capita per day. Despite increased food production, there has been no change over the past two decades in the estimated 25 percent of the population who are undernourished (FAOSTAT 2012). The National Nutrition Survey 2011 (Bhutta 2012) reports that 57 percent of the population does not have food security.

Water and agricultural production depend on managing these many forms of resource variability and uncertainty. An overarching pattern in the issues presented above is that while the Indus Basin is richly endowed with land and water resources vital for the agricultural economy, it faces high levels of variability and uncertainty in climate, hydrology, agricultural sustainability, food consumption, and natural hazards.

Study Objective

The objective of this study is to assess the impacts of climate risks and various development alternatives on water and agriculture in the Indus Basin of Pakistan. The study analyzes inter-relationships among the climate, water, and agriculture sectors of the country. A better understanding of how these sectors are linked will help plan future investments in these sectors. Many different forums and policy reports in Pakistan recognize the important role of water management in the productivity of the agriculture sector and overall food security. However, the factors affecting the availability and use of water and its connection to the factors mentioned above are not always comprehensively addressed (with systems-based models) in federal and provincial planning documents and budgeting. This study provides a systems modeling framework for these purposes.

Analytically, this objective is achieved by integrating several different modeling environments, including a model of Upper Indus snow and ice hydrology (critical for determining overall water availability), an agro-economic optimization model of the IBIS, and an updated computable general equilibrium (CGE) model of Pakistan's wider macro-economy. This integration of models helps to frame recommendations for strengthening water, climate, and agriculture planning, policies, and research priorities for the Indus Basin.

The key climate risk challenges to be examined in the context of this modeling framework include: (1) limited water storage, (2) problematic trends in surface water and groundwater use, (3) inflexible and uncertain water allocation institutions, and (4) low water-use efficiencies and productivity.

Policies and Plans

The Indus Basin of Pakistan, like other complex river basins, faces a common set of institutional and policy challenges: (1) international treaty tensions over upstream development; (2) sectoral integration across water, agriculture, environment, climate, and energy agencies at the national level; (3) national-provincial coordination in a federal system of government; and (4) interprovincial water conflict resolution. The Government of Pakistan has responded to these tensions in several creative ways, beginning with the establishment of the Water and Power Development Authority (WAPDA) as a semiautonomous federal water agency, followed by the Indus River System Authority (IRSA) as a federal-provincial coordinating organization. The Planning Commission's *Vision 2030* (GPPC 2007) and reports of the Intergovernmental Task Force on Climate Change (GPPC 2010) and Task Force on Food Security (GPPC 2009) articulate integrative analyses and recommendations. While a comparably broad national water policy is needed, reform of provincial irrigation institutions has been initiated over the past decade in Punjab and Sindh provinces. Notwithstanding earlier efforts at integrative planning, much more institutional coordination, integration, and conflict transformation will be required to address the substantive water, climate, and agriculture issues highlighted.

This common set of water and agricultural policy challenges is complicated by several dynamic stresses and institutional shifts in the Indus Basin. First, a series of catastrophic floods, droughts, and earthquakes over the past decade have led to the establishment of new, but not yet adequate, national and provincial disaster management agencies. Second, growing concerns about climate change led in 2012 to the adoption of a national climate change policy and creation of a Ministry of Climate Change. These new developments are promising but require major capacity-building and rigorous coordination with established water, power, and agricultural agencies.

Furthermore, these structural pressures and responses are occurring within the context of major constitutional devolution from the national to the provincial level. The 18th amendment to the constitution, passed in April 2010, eliminated the concurrent list of federal and provincial responsibilities and devolved most of the functions on that list to the provincial level. This constitutional change requires stronger policy links between the federal water sector and the provincial irrigation and agricultural sectors, which will require both vision and budget support at the federal and provincial levels. Moreover, assessments of climate impacts and adaptations (as they relate to the agriculture and water sectors) must devote increased emphasis to provincial planning, management, and governance.

Most national and provincial development plans continue to focus on the role of infrastructure in addressing challenges of water and agriculture. However, recent documents highlight the increasing importance of improving irrigation efficiency and reducing fiscal shortfalls in the irrigation revenue system. Moreover, increasing yields is identified as a key area of needed improvement, particularly as it relates to adaptability to climate change (for example, accelerated crop breeding and adoption of genetically modified Bt cotton).

Various forums and policy reports increasingly recognize the important role that water management plays in agricultural productivity; however, this relationship is not always comprehensively addressed (with systems-based models) in federal and provincial planning documents and budgets. This discrepancy underscores the need to address sector gaps and reduce uncertainties among the irrigation, agricultural, and climate change policies for Indus Basin management.

Upper Indus Basin Hydrology and Glaciers

While it is generally agreed that a significant percentage of the Indus River flow originates in the mountain headwaters of the Karakoram Himalaya, western Himalaya, and Hindu Kush Mountains, there is no consensus regarding the role or importance of this runoff for the complex hydrometeorological environments that characterize the mountain catchments. In particular, there has been considerable speculation concerning the importance of glaciers in the flow volume and timing of the Indus River and its tributaries, as well as on the potential impact of climate change on this water supply. Although these concerns are recognized, few analyses have described the role of glaciers in the hydrologic regime of these mountains, in large part due to the inaccessibility and altitude (4,000–7,000 meters [m]) of Himalayan glaciers.

Estimates of the potential impact of a continued retreat of the glaciers are derived on the basis of use of disaggregated low-altitude databases, topography derived from satellite imagery, and simple process models of water and energy exchange in mountain regions. The surface area of the Upper Indus Basin (UIB) is approximately 220,000 square kilometers (km²). Of this surface area, more than 60,000 km^2 is above 5000 m, the estimated mean altitude of the summer-season freezing level. It is assumed that significant melt of glaciers does not occur over most of this upper zone. The glaciers of the region flowing outward from this zone have been estimated to have a surface area of approximately 20,000 km^2, of which 7,000–8,000 km^2 is below the summer-season freezing level. It is this 7,000–8,000 km^2 area, in the ablation zone, that is the source of the bulk of the annual glacier melt water flowing onto the Indus River tributaries.

The two principal sources of runoff from the UIB are winter precipitation, as snow that melts the following summer, and glacier melt. Winter precipitation is most important to the seasonal snow runoff volume, while summer temperature contributes most to glacier melt volume. Drawing a clear distinction between the runoff volumes resulting from snow melt and glacier melt is difficult. The primary zone of melt water from both sources is maximized at around 4,000–5,000 m, as a result of a combination of maximum terrain surface area, maximum glacier surface area, and maximum snow water equivalent deposition occurring there. This is the altitudinal zone generally reached by the upward migration of the freezing level during the months of July and August, which is also the time of maximum runoff.

Using a simple model of these dynamics, it is estimated that glacier runoff contributes approximately 19.6 million acre-feet (MAF) to the total annual flow of the UIB: 14.1 MAF from the Karakoram Himalaya, 2.3 MAF from the western Himalaya, and 3.2 MAF from the Hindu Kush. This represents an estimated 18 percent of the total flow of 110 MAF from the mountain headwaters of the Indus River. Thus, the most probable source for a majority of the remaining 82 percent is melt water from the winter snowpack (figure ES.1). Approximately

Figure ES.1 Contributions of Snowmelt and Ice Melt to Total Runoff for Sub-Basins in the Upper Indus Basin

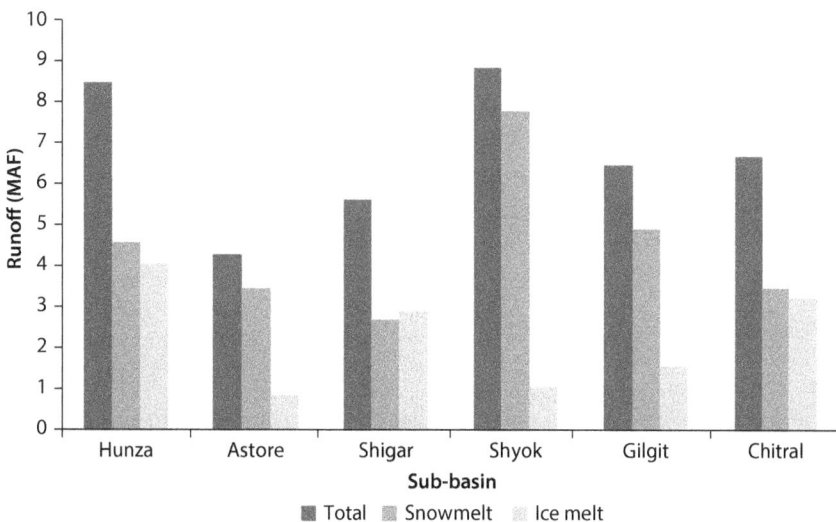

Note: MAF = million acre-feet.

80 percent of the annual stream flow volume in the tributaries of the UIB occurs during the summer months of mid-June to mid-September.

In order to assess the potential impact of climate change scenarios on stream flow in the UIB, it is useful to distinguish between changes that may result from variations in precipitation and those related to changes in temperature. The volume of runoff from winter snowmelt will be determined primarily by variations in winter precipitation. On the other hand, glacier melt water production will vary with the energy availability (changes in temperature, primarily during summer) during the melt season at the glacier surface.

Because of the complexities at these high elevations, general circulation models (GCMs) are unlikely to have much value for forecasting purposes. Therefore, a major investment is needed in snow and ice hydrology monitoring stations, further scientific research, and forecasting to improve the hydrologic predictability of the UIB.

Future Climate Scenarios

Analysis of GCM outputs supports the subsequent modeling approach in which ranges of climate risks are informed by GCM outputs but not driven directly by them. Moreover, given the uncertainties these outputs demonstrate, a wider and more extreme range of possible climate futures is considered.

Historically, over the entire country, the mean temperature has increased by about 0.6°C over the last century. Annual precipitation has also increased by about 25 percent (63 millimeters [mm]) over the same period. Precipitation patterns across provinces and within the year are less clear. Each river in the Indus system has its own hydrologic regime (for example, Chenab, Indus, and Jhelum). In general, the Indus system exhibits less variability in a year than other major river basins in the world (for example, the Ganges). In part this is a reflection of the moderating impact that snow and ice play in the overall hydrology.

The general findings from a wide range of GCM outputs (and emissions scenarios) show agreement among models regarding continued increases in temperature into the future. Increases are estimated to be at worst close to 3°C by the 2050s. The temperature increases in both summer and winter are higher in northern Pakistan than in southern Pakistan. Moreover, temperature increases tend to be on average higher during winter than summer. The models do not agree on changes in precipitation because standard errors are large; however, there is some indication of a general trend in increased precipitation during the summer and a decrease during the winter. The changes appear to be more pronounced in the southern parts of the country. These models are likely to be more reliable for the irrigated plains than for the mountainous upper basin.

Given the orographic complexity of the UIB, future projections of volumes entering the Indus system are inconclusive. Moreover, using the snow and ice hydrology model developed in this study and a wide range of climate futures, the postulated impact of climate change on inter-annual stream flow variations is generally comparable with the current inter-annual variations. Therefore, it is anticipated that for the UIB, the primary impact of all but the most extreme climate change scenarios could be a shift in the timing of peak runoff, and not a major change in annual volume.

Modeling Water, Climate, Food, and the Economy

The modeling framework developed in this study integrates two models in addition to the snow and ice hydrology model. The first model is an agro-economic optimization model that takes a variety of inputs (for example, agronomic, irrigation system data, water inputs) to generate the optimal crop production across the provinces (subject to a variety of physical and political constraints) in the existing IBIS for every month of the year. The objective function for this model is primarily the sum of consumer and producer surpluses. The second model is a CGE model for the Pakistan macro-economy. This integration helps to better understand how changes in climate risks impact the macro-economy and different household groups through the agriculture sector.

Sensitivity analyses indicate that the objective value is most sensitive to stream inflow, crop water requirement, and depth to groundwater parameters. The objective value decreases to almost 60 percent of the baseline when the inflows drop to its 90 percent exceedance level (101 MAF). When the inflow increases to its 10 percent exceedance level (209 MAF), the objective value shows no significant change because the system is unable to generate more economic benefits in the basin. This may be due to crop productivity limits, policy constraints on water allocation, or physical constraints. Increasing crop water requirements, which are proportional to air temperature in the study's analysis, also results in a substantive decrease in the objective function; for example, about a 40 percent reduction occurs with a crop water requirement increase of 35 percent (corresponding to a 6.5°C increase) of the baseline value.

Meanwhile, the 1991 Indus Water Accord ("the Accord"), which determines the provincial allocations of the Indus, is a critical constraint in the system. The difference in objective value is a factor of 2 between strict adherence to the Accord and relaxation of the Accord. According to this study model results, if the Accord is relaxed and water is allocated economically and optimally within and among provinces, both Punjab and Sindh could benefit. The system-wide net revenue will increase by about PRs 158 billion (almost US$2 billion)—with PRs 83 billion additional in Punjab and PRs 82 billion additional in Sindh. Moreover, by relaxing the Accord and implementing economically based water allocation mechanisms, provinces will be better able to manage extreme events (for example, drought) by more reliably meeting system-wide demands. However, this effort would need to be supported by investment in effective, transparent, real-time water delivery measurement systems; capacity-building in IRSA and WAPDA for technical decision-support systems and forecasting; and equally substantial investment in trust-building among stakeholders. Finally, even though it is unlikely and probably unwise that the Accord constraint should—by itself—be relaxed, there is room for flexible policy adjustments and mechanisms within the wider framework of the present Accord (for example, interprovincial exchange of surplus allocations, water banking, and leasing arrangements),[1] as well as for improved water allocation within provinces, which the modeling results suggest should be pursued on agro-economic grounds.

Climate Risk Scenarios

Climate change projections show great uncertainty and questionable skill in this region. To generate a wider range of potential climate scenarios, the study team used combinations of corresponding inflow and crop water requirement parameters. Inflow was varied from 10 to 90 percent exceedance probability, and the crop water requirement was varied to correspond to a 1°–4.5°C temperature increase (possibly occurring around the 2020s and 2080s, respectively). Furthermore, since much of the waters in the system originate in the UIB in the Himalayas, climate change impacts (using corresponding temperature and precipitation changes) on snow and ice in the UIB, and ultimately on the inflows into the Indus main-stem basin, were considered. These climate futures represent a plausible range of climate changes within the next 80 years, consistent with recent observations and theory.

Generally, negative impacts are estimated under these climate risk scenarios. GDP, Ag-GDP, and household income are estimated to decrease on average by 1.1, 5.1, and 2.0 percent, respectively, on an annual basis (figure ES.2a). In the most extreme future scenario—when inflow is at 90 percent exceedance probability and the temperature increases 4.5°C—GDP, Ag-GDP, and household income are estimated to decrease annually by 2.7, 12.0, and 5.5 percent, respectively. Most of the negative impacts on incomes will occur for those households outside of the agriculture sector (except for those living in provinces other than Punjab and Sindh) that would be faced with an increase in food prices (figure ES.3). Since the increase in prices is larger than the decrease in production, farm-related households will likely benefit. However, nonfarm households in towns and cities will have to pay more for food, resulting in decreased household incomes.

Total crop production is estimated to decrease up to 13 percent (figure ES.2b). The change in hydropower generation varies the most, from a 22 percent increase to a 34 percent decrease. Increases are a result of more surface water becoming available from more snowmelt. Impacts are greatest for crop production in Sindh, at around 10 percent on average (figure ES.4). Irrigated rice, sugarcane, cotton, and wheat demonstrated the greatest sensitivity to climate, and changes represent both response to climate and dynamic responses to water availability and price changes.

In order to assess the likelihood of low probability but high impact climate changes ("surprises"), possible "worst" and "best" case climate futures were evaluated. The worst case was defined as 90 percent exceedance inflow, a forward monthly hydrograph shift, 20 percent less rainfall, 20 percent more water

Figure ES.2 CGE and IBMR Economic Outcomes under Climate Risk Scenarios

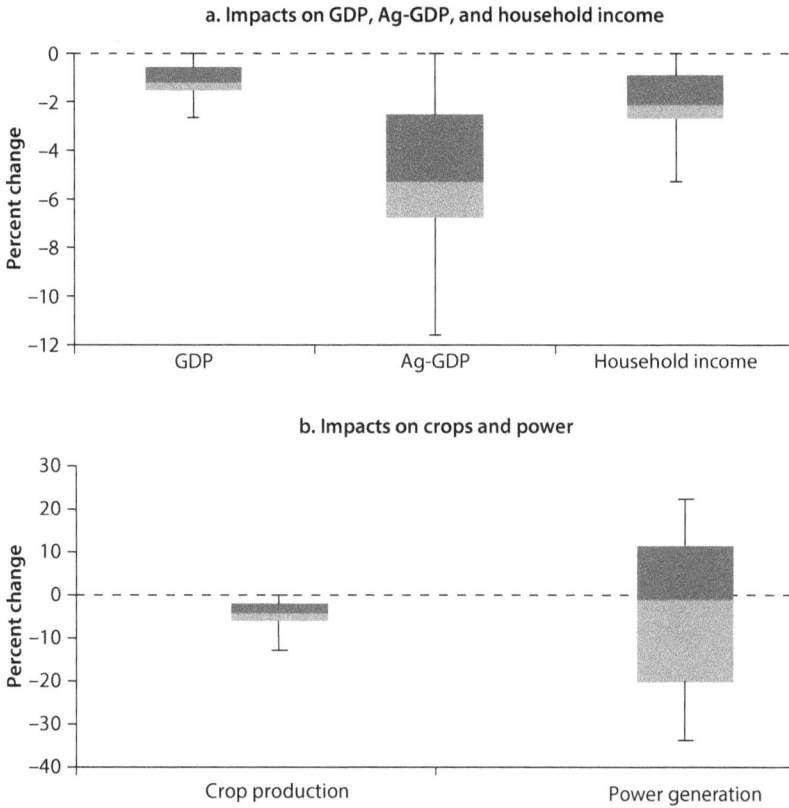

a. Impacts on GDP, Ag-GDP, and household income

b. Impacts on crops and power

Note: CGE = computable general equilibrium, IBMR = Indus Basin Model Revised.

Figure ES.3 Different Household Income Changes under Climate Risk Scenarios

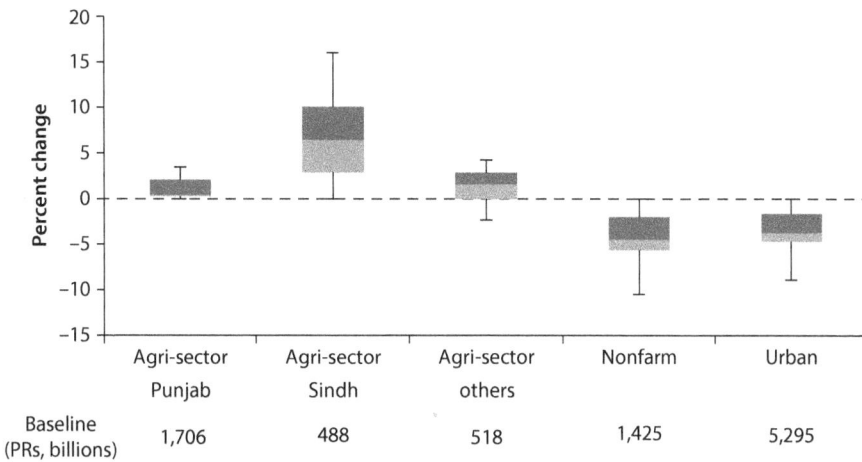

	Agri-sector Punjab	Agri-sector Sindh	Agri-sector others	Nonfarm	Urban
Baseline (PRs, billions)	1,706	488	518	1,425	5,295

Figure ES.4 Crop Production Changes under Climate Risk Scenarios

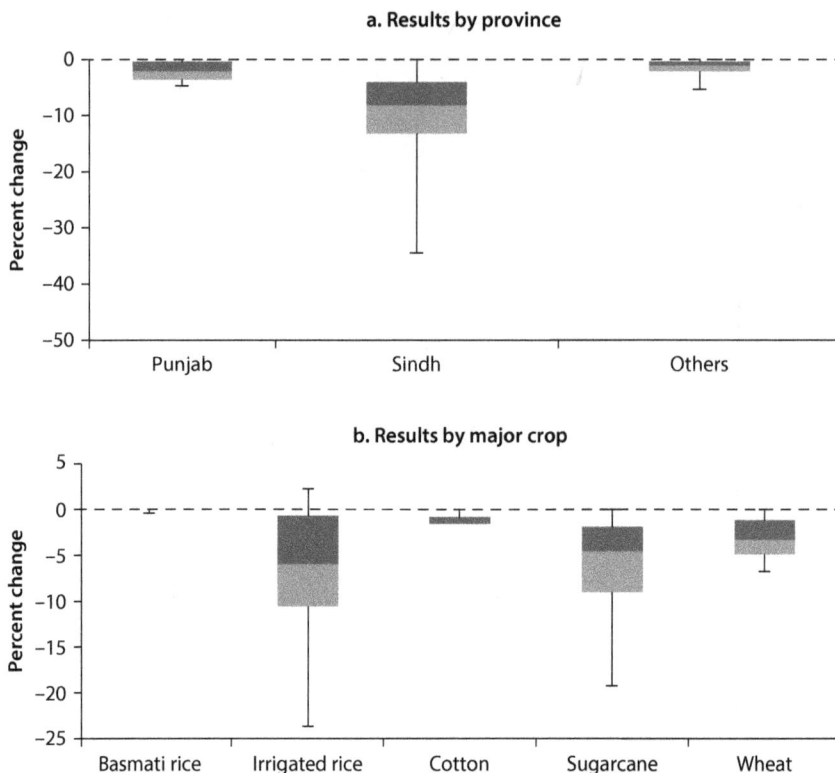

a. Results by province

b. Results by major crop

requirement (consistent with a +4.5°C change), and groundwater table depths that are 20 percent deeper throughout the basin. The best case was defined as the 10 percent exceedance inflow, 20 percent more rainfall, no change in the existing crop water requirements, and groundwater table depths that are 20 percent shallower. In the "worst" case, GDP, Ag-GDP, and household income decrease by 3.1, 13.3, and 6.7 percent, respectively, on an annual basis. In the best case, GDP, Ag-GDP, and household income increase by 1.0, 4.2, and 1.3 percent, respectively. These results represent a wide range of economic futures given current conditions, including the current Accord allocation, in this basin.

Adaptation Investment Scenarios

Three possible adaptation investments were evaluated: (1) canal and watercourse efficiency improvements (CANEFF) to bring the system to 50 percent system-wide efficiency levels; (2) construction of new reservoirs to introduce an additional 13 MAF (NEWDAM); and (3) investments in agricultural technologies to increase crop yield (CYIELD) by 20 percent. To examine the role that these investments play over time, the original Indus Basin Model Revised-2012 (IBMR-2012) was modified for multiyear analysis.

Figure ES.5 Cumulative Distribution Functions of IBMR-2012 Objective Value for Different Adaptation Investments (without Climate Risk Scenarios)

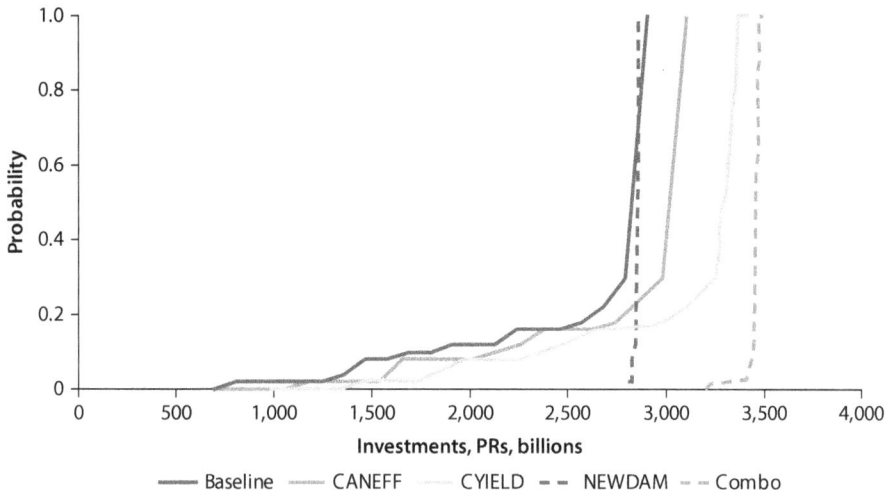

Note: CANEFF = canal and watercourse efficiency improvements, CYIELD = crop yield, NEWDAM = construction of new reservoirs. The cumulative distribution function is a graph of the value of the objective function versus the probability that value will occur.

Figure ES.5 shows the cumulative distribution functions (CDFs) of the objective function for each of these investment scenarios. The CDF is a graph of the value of the objective function versus the probability of that value (or less) occurring. The CANEFF and CYIELD investments shift the CDF to the right of the baseline, indicating that the objective value tends to increase under these investments. The long left-side tails of the CDFs of these two investments are a result of the very low values that occur in difficult years, such as droughts. The NEWDAM investment is unique in that it eliminates the left-side tail, showing that additional storage reduces the probability of very low objective values, thus mitigating the effects of drought years. It does not, however, increase the objective value under normal and high flow years, primarily because the objective function does not include the economic benefits from additional hydropower generation. Thus, this analysis pertains to the value of storage for irrigated agriculture but not for other services such as hydropower and flood risk reduction. In addition, while the increased reservoir volume may supply more water, due to the constraints of the Accord, that water cannot be put to use effectively.

Canal efficiency and crop yield investments show a potential to minimize the impacts of future climate risks and increase food production. Investments in canal and watercourse efficiency and crop technologies are estimated to increase average crop production by 5–11 percent annually (figure ES.6a). Thus, these investments will have positive impacts on the macro-economy and households. Instead of losses of 1.1, 5.1, and 2.0 percent of GDP, Ag-GDP, and household income, respectively, estimated under climate change scenarios, with these

Figure ES.6 Sector Outcomes for Adaptation Investments under Climate Change Conditions

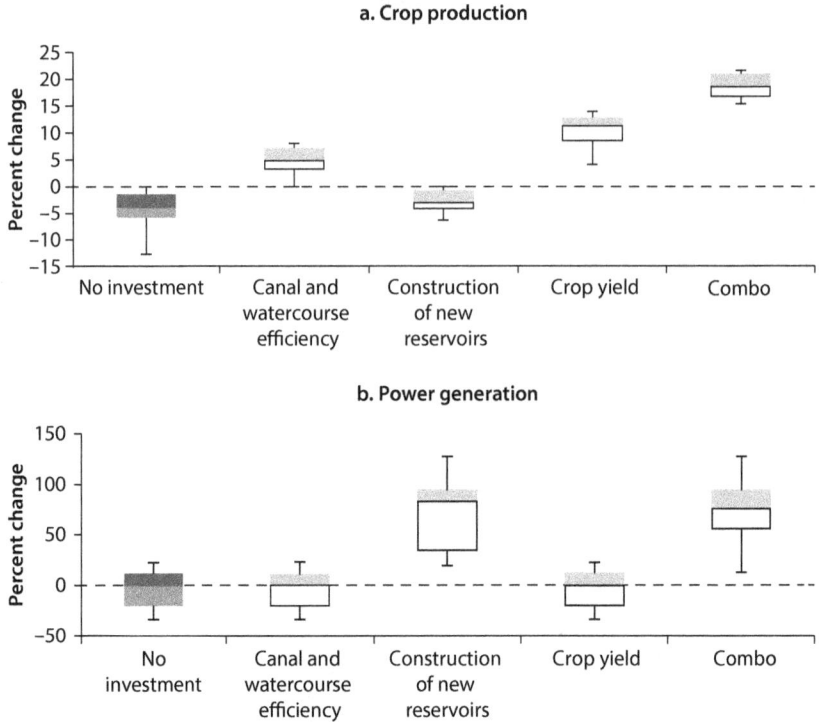

a. Crop production

b. Power generation

Note: CANEFF = canal and watercourse efficiency improvements, CYIELD = crop yield, NEWDAM = construction of new reservoirs.

Table ES.1 Impact of Different Adaptation Investments under Climate Risk Scenarios

	GDP	Ag-GDP	Household income
Average change without investments (%)			
No investment	−1.1	−5.1	−2.0
Average gain with investments (%)			
CANEFF	2.04	9.32	3.21
NEWDAM	0.29	1.50	0.64
CYIELD	3.66	16.70	5.42
Combo	6.05	27.40	7.45

Note: CANEFF = canal and watercourse efficiency improvements, CYIELD = crop yield, NEWDAM = construction of new reservoirs.

adaptation investments, gains can still be realized (see table ES.1). For example, the average Ag-GDP will decrease by 5.1 percent without any adaptation but will increase by 4.2 and 11.6 percent with the CANEFF and CYIELD investments, respectively. These investments are still vulnerable under low-flow drought conditions. On the other hand, investment in additional storage (NEWDAM) reduces the impacts of climate change but does not completely mitigate them. In addition, the NEWDAM investment would also realize quite large hydropower generation benefits (figure ES.6b) despite climate changes.

Table ES.2 Protein and Carbohydrate Supply and Requirement under Climate Change Estimates

	Population (millions)	Cereal-based protein and carbohydrate demand (tons, millions)	Protein and carbohydrate supply (tons, millions)			
			Baseline	CANEFF	NEWDAM	CYIELD
Baseline	167.4	10.1	16.3	18.0	16.4	19.8
2020-low P	227.8	13.7	16.1	17.7	16.2	19.4
2020-high P			16.2	17.8	16.3	19.5
2050-low P	307.2	18.4	15.8	17.2	15.9	19.0
2050-high P			15.9	17.4	15.9	19.1
2080-low P	386.7	23.1	15.5	16.8	15.6	18.6
2080-high P			15.5	16.8	15.6	18.7

Note: CANEFF = canal and watercourse efficiency improvements, CYIELD = crop yield, NEWDAM = construction of new reservoirs. Shaded cells indicate supply is less than demand. Baseline means current climate condition. Low and high P represent lower and higher precipitation projections from the GCMs.

Changes in crop production are directly related to the achievement of food self-sufficiency. The supply and demand of wheat was used to evaluate this issue. The protein and carbohydrate demands in the 2020s, 2050s, and 2080s were estimated based on future population estimates (table ES.2). The supply is projected to be less than the demand by the 2050s without any investment intervention. Only the CYIELD investment can maintain the necessary production to meet future protein and carbohydrate requirements. By the 2080s, none of the investments will be able to supply sufficient protein and carbohydrate requirements for the country.

Disaggregating these findings by province shows that Punjab will be able to meet its protein and carbohydrate demands, even out to 2080. The real food self-sufficiency challenge will be in Sindh, even as early as 2020. Note that it is assumed that interprovincial trading does not change. Also, the evaluation is in terms of self-sufficiency and does not evaluate changes in the ability to import food or potential changes in diet.

Without specific interventions, environmental considerations such as flow to the sea, changes in depth to groundwater, and the overall salinity situation, are projected to worsen. In particular, groundwater depletion in the fresh groundwater area and the basin-wide salinity issue will become worse if no policy intervention is made. The study analysis reveals that the decreases in net benefits (as a percentage of the baseline) are not that significant under a scenario where safe groundwater yields are enforced. However, without intervention, the long-term trends are troubling. The net recharge in fresh groundwater areas is negative in all provinces, with the largest values estimated in Punjab, suggesting continued declining water tables. This decline also contributes to increased saline water intrusion. Additionally, salt accumulation is positive in all provinces and in both fresh and saline areas. Given the scale of these issues, a new phase of truly visionary planning is needed for conjunctive management of surface and groundwater.

Conclusions

This analysis identified, first, the key hydroclimatic sensitivities and robust aspects of the IBIS. Second, the models used here are among the best mathematical representations available of the physical and economic responses to these exogenous future climate risks. However, uncertainty exists because, as in all modeling approaches, parameters may not be known with precision and functional forms may not be fully accurate. Thus, careful sensitivity analysis and an understanding of and appreciation for the limitations of these models are required. If undertaken, further collection and analysis of critical input and output observations (for example, snow and ice data), as well as practical measures for improving productivity under a changing climate, would enhance this integrated framework methodology and future climate impact assessments.

The precise impact of climate risks on the Indus Basin remains to be seen. This much is known, however: Climate change will pose additional risks to Pakistan's efforts to meet its water and food self-sufficiency goals, goals that are key to reducing poverty, promoting livelihoods, and developing sustainably. As its population grows, Pakistan's ability to meet basic food requirements and effectively manage water resources will be critical for sustaining long-term economic growth and rectifying widespread food insecurity and nutrition deficiencies. These are challenges above and beyond what Pakistan is already facing, as evidenced by the extreme hydrologic events of 2009–11. Strategic prioritization and improved planning and management of existing assets and budget resources are critical. These strategic choices will be largely dependent on a sound assessment of the economics of these impacts.

The integrated systems framework used in this analysis provides a broad and unique approach to estimating the hydrologic and crop impacts of climate change risks, the macro-economic and household-level survey responses, and an effective method for assessing a variety of adaptation investments and policies. In assessing the impacts, several different modeling environments must be integrated to provide a more nuanced and complete picture of how water and agriculture interrelate. Moreover, such a framework allows for extensive scenario analysis to identify and understand key sensitivities. This analysis is critical to making decisions in a highly uncertain future. Finally, through this integration of multiple disciplines, a richer and more robust set of adaptation investment options and policies for the agriculture and water sectors can be identified and tested. Continued refinements to the assessment approach developed in this volume will further help to sharpen critical policies and interventions by the Pakistan Government.

Note

1. These types of approaches can be seen increasingly in basins such as the Colorado River in the United States.

References

ADB (Asian Development Bank) and World Bank. 2010. "Pakistan Floods 2010 Damage and Needs Assessment." Pakistan Development Forum, Islamabad.

Bhutta, Z. 2012. *Pakistan—National Nutrition Survey 2011*. Karachi, Pakistan: Aga Khan University Pakistan Medical Research Council Nutrition Wing, and Ministry of Health.

FAOSTAT (Food and Agricultural Organization of the United Nations Statistical Database). 2012. Database of Food and Agriculture Organization of the United Nations, Rome. http://faostat.fao.org.

GPPC (Government of Pakistan, Planning Commission). 2007. "Agricultural Growth: Food, Water and Land." In *Vision 2030*, GPPC, 51–60. Islamabad. http://www.pc.gov.pk/vision2030/Pak21stcentury/vision%202030-Full.pdf.

———. 2009. "Final Report of the Task Force on Food Security." Islamabad.

———. 2010. "Task Force on Climate Change Final Report." Islamabad.

CHAPTER 1

Two Years in the Life of the Indus River Basin

This study was undertaken at a pivotal time in the region. The weak summer monsoon in 2009 created drought conditions throughout the country. This followed an already tenuous situation for many rural households faced with high fuel and fertilizer costs and the impacts of rising global food prices. To make matters worse, catastrophic monsoon flooding in 2010 affected more than 20 million people, their housing, infrastructure, and crops. Damages from this single flood were estimated at around US$10 billion (ADB and World Bank 2010), with about half attributed to losses in the agriculture sector. Whether such observed extremes were evidence of climate change and the extent to which the country is resilient to these shocks were the questions these events raised. It is thus timely, if not critical, to focus on climate risks for water, agriculture, and food security in the Indus Basin.

Background and Problem Statement

The Indus Basin has an ancient and dynamic record of irrigation development and change. Settlements of the Indus valley's Harappa civilization date back some five millennia. Traces survive of inundation channels that flowed across the floodplains during the monsoon season, enabling flood farming of fuel, fodder, and small grain crops within the riparian corridor. Large check dams known as *gabarbands* impounded water on hill torrents and tributary watersheds. Sophisticated urban sewage systems and baths served cities like Mohenjo-Daro in the lower Indus valley. These cities and smaller settlements were abandoned in the second millennium BC, by some accounts due to flooding, salinity, and river channel change (Giosan et al. 2012; Wright 2010). In one major drainage on the arid eastern side of the middle Indus valley, the Ghaggar-Hakra river channel shifted course in the Harappan era, leading to the abandonment of hundreds of settlements. These historical events invite questions about long-term sustainability in the context of dynamic hydroclimatic variability (Mughal 1997).

Localized irrigation flourished again during the medieval period. Innumerable shallow, hand-dug, masonry-lined wells provided water for local irrigation agriculture and livestock husbandry. Water buckets were lifted by ropes, pulleys, and Persian wheels (geared mechanisms that lifted chains of terracotta water pots) powered by humans and draft animals. In Balochistan, deeper wells tapped into hillside groundwater supplies, and tunnels known as *qanats* conveyed water to irrigated fields and settlements. These local groundwater systems were succeeded by a vast surface water canal irrigation system diverted by long masonry-clad barrages across the Indus and its major tributaries from the mid-19th to late 20th century. The benefits of dramatically expanded irrigated acreage and production were offset in some areas by seepage, waterlogging, salinity, and depleted environmental flows. Development of tubewell pumping technology in the mid-20th century improved the flexibility of irrigation and groundwater management but brought its own issues of unregulated withdrawals and secondary soil salinization. At the start of the 21st century, the core challenge was to achieve dramatically higher productivity through improved management of soil moisture, groundwater, canal irrigation, and environmental flows in ways that are adaptive and resilient.

Pakistan relies on the largest contiguous irrigation system in the world, known as the Indus Basin Irrigation System (IBIS), providing basic food security and water supply for all sectors of the economy (map 1.1). The basin that supports this irrigation system comprises the Indus River main stem and its major tributaries—the Kabul, Jhelum, Chenab, Ravi, and Sutlej rivers. The IBIS has 3 major multipurpose storage reservoirs, 19 barrages, 12 inter-river link canals, 45 major irrigation canal commands (covering over 18 million hectares), and over 120,000 watercourses delivering water to farms and other productive uses. Annual river flows are about 146 million acre-feet (MAF), of which about 106 MAF of water is diverted from the river system to canals annually (COMSATS 2003). The total length of the canals is about 60,000 km, with communal watercourses, farm channels, and field ditches running another 1.8 million km. These canals operate in tandem with a vast and growing process of groundwater extraction from private tubewells.

The IBIS is the backbone of the country's agricultural economy. The agriculture sector supported by this system plays a critical role in the national economy and the livelihoods of rural communities. Agriculture contributes some 22 percent to Pakistan's gross domestic product (GDP), down from 27 percent in 1989 and 46 percent in 1960, due primarily to more rapid growth in the services sector; 45 percent of the labor force is employed in the agriculture sector. The value of agricultural production continues to grow at an average annual rate of approximately 3 percent (figure 1.1a). However, the inter-annual variability of agricultural value added to GDP is high (figure 1.1b), demonstrating existing vulnerabilities to climate risks.

The largest crop by tonnage is sugarcane, followed by wheat, milk, rice, and cotton (FAOSTAT 2012). In terms of economic value, milk tops the list, followed by wheat, cotton, rice, meat, and sugarcane. These patterns indicate

Map 1.1 Indus Basin Irrigation System

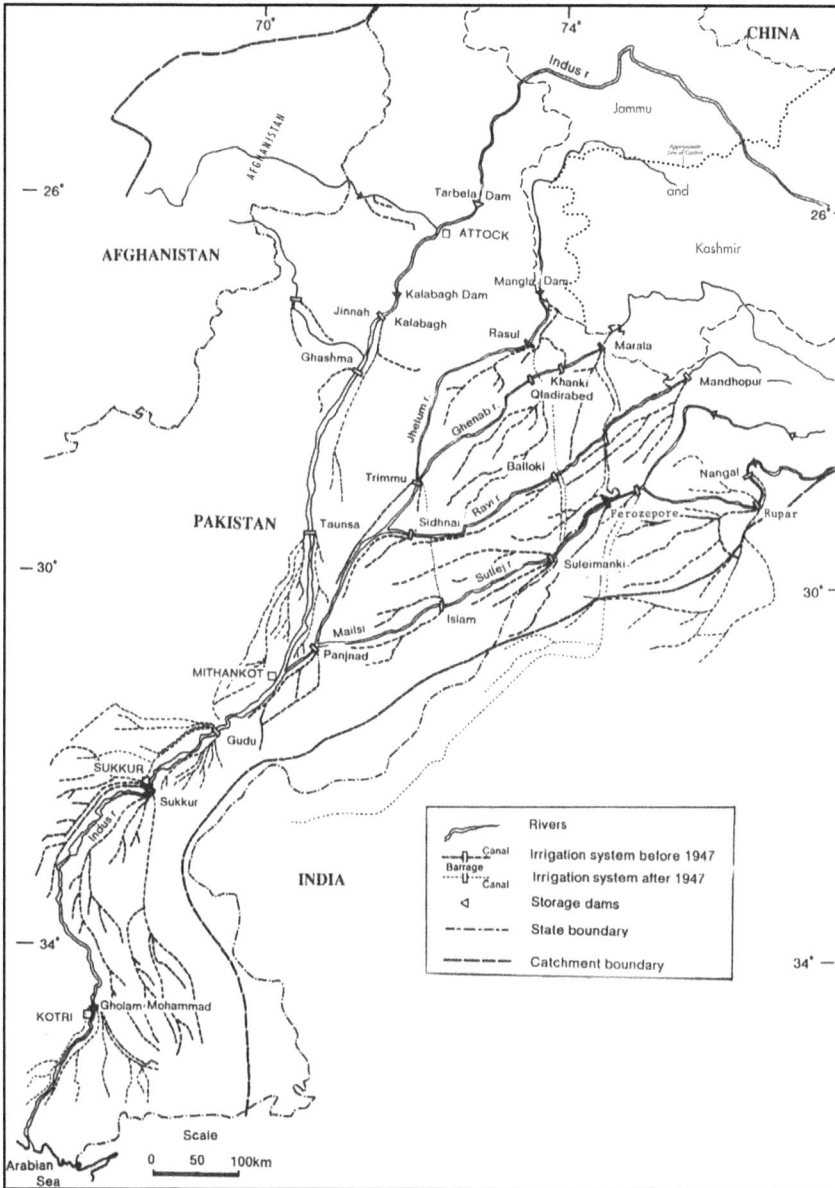

the rising economic significance of dairy and livestock products. Some 64 percent of Pakistan's population is rural, and an estimated 40–47 percent of the labor force is involved in agriculture (World Bank 2012b). Women constitute an increasing proportion of the agricultural labor force, at 30 percent, double the proportion of 20 years ago (FAOSTAT 2012). Agricultural

Figure 1.1 Value and Growth of Agricultural Production

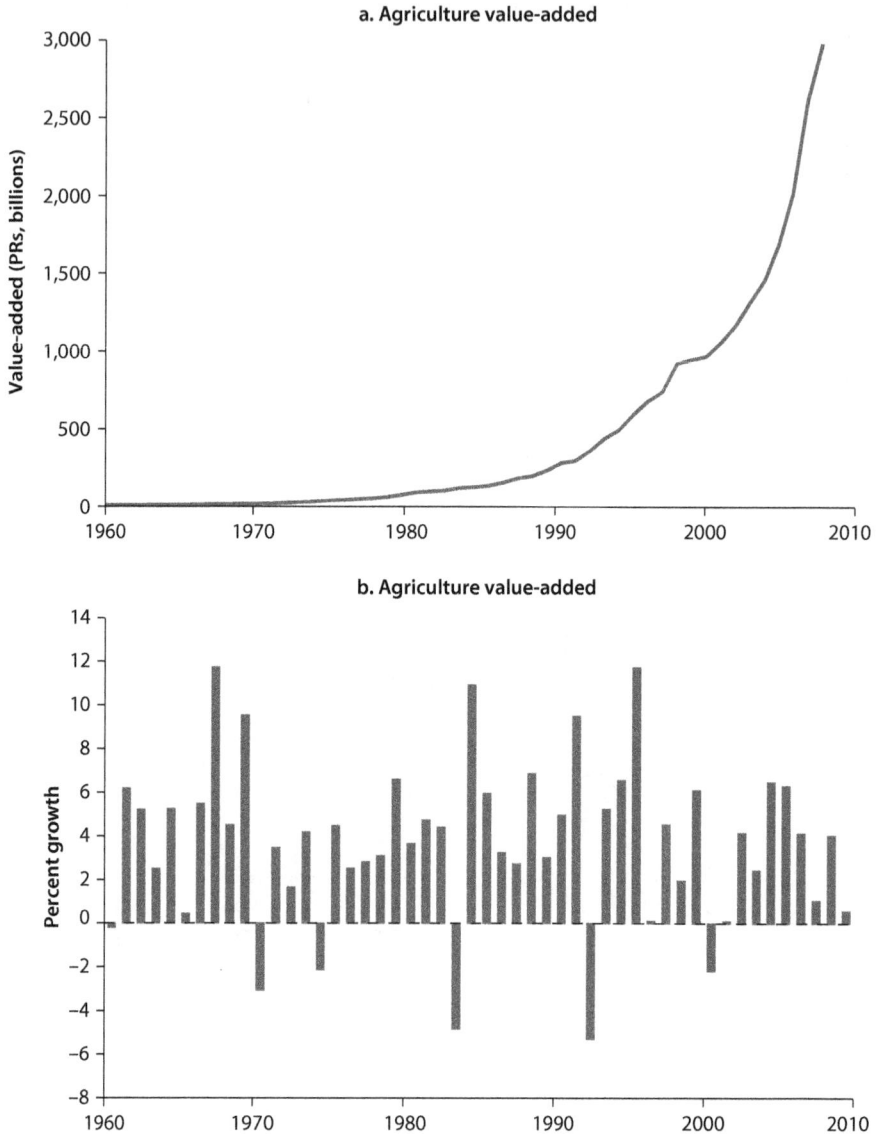

a. Agriculture value-added

b. Agriculture value-added

Source: World Development Indicators 2012.

mechanization has also increased at a rapid rate; tractors have completely replaced draft animal power, and new technologies of precision land leveling and drip irrigation have expanded.

Irrigated land supplies more than 90 percent of agriculture production. Agriculture in most areas is not possible without irrigation because the climate of Pakistan is arid to semi-arid, with low and variable rainfall. Only 28–35 percent of the total land area is arable, and that proportion has not increased significantly in recent decades. However, the irrigated portion of arable land has grown over the past decade (from about 65 percent in 2001 to almost 75 percent in 2009), which has contributed to increased agricultural production and yields. Rain fed (*barani*) crops, with much lower and less reliable yields than irrigated crops, nevertheless are increasingly important but are highly vulnerable to climate variability. Annual rainfall over much of the lower basin is not more than 150 millimeters (mm) per annum, with high potential evapotranspiration rates, ranging from 1,250 to 2,800 mm per annum. However, a substantial amount of water flows into the Indus Basin, which drains 70 percent of the country (566,000 km^2).

The rivers of the Indus Basin have glaciated headwaters and snowfields that, along with monsoon runoff and groundwater aquifers, provide the major sources of water for Pakistan. Currently, about 50–80 percent of the total average river flows in the Indus system are fed by snow and glacier melt in the Hindu-Kush-Karakoram (HKK) part of the Himalayas, with the remainder coming from monsoon rain on the plains. There are more than 5,000 glaciers covering about 13,000 square kilometers (km^2) in the Upper Indus river basin catchment (map 1.2).

Map 1.2 Glaciers and Drainage Area in Upper Indus Basin, Pakistan

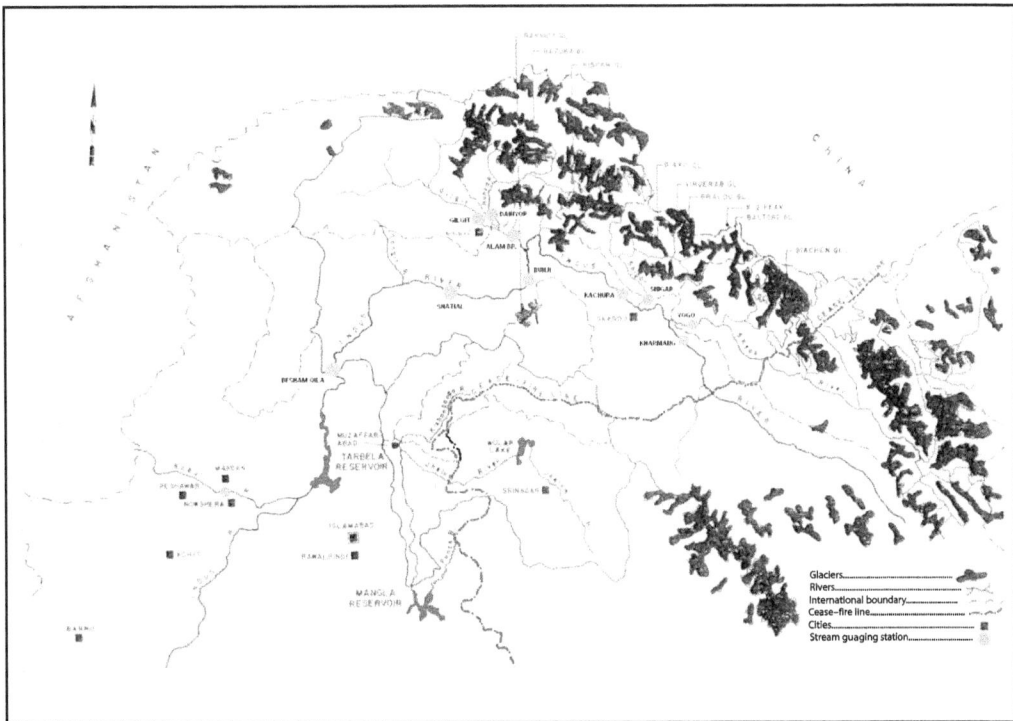

The supply of water stored in glaciers and snow is projected to decline globally during the 21st century. However, the patterns of depletion and accumulation vary regionally and locally. Some glaciers in the Upper Indus are increasing in depth and size, in contrast with the more general (but still variable) pattern of glacial retreat in the Himalayan range to the east. However, the bulk of the melt waters in the region come more from snow fields than glaciers (see chapter 3). In part because of this complex mix of sources, the variability observed in the Indus is not as large as for other major rivers in the world (for example, the Ganges). Variability in the distribution and timing of snowfall and changes in the melting of snow and ice, however, may be amplified by climate change and have implications for managing basin water resources.

Monsoon rainfall contributes to flood hazards in highly variable ways. The remainder of the water availability after melts is from the annual monsoon system. This contribution is even more variable than that of Upper Basin inflows. Monsoon floods have displaced hundreds of thousands of people in Pakistan (in 2003, 2005, 2008, 2010, and 2011) in the last decade alone (Brakenridge 2012). The same decade witnessed a severe multiyear drought. Finally, changes in temperature, precipitation, and atmospheric CO_2 concentrations have a direct impact on agricultural yields. Such changes, in addition to climate risks that the country already faces, pose major challenges for water managers over the coming 20–30 years.

Managing groundwater resources continued to be a major challenge in the Indus Basin. Waterlogging and salinity have been major concerns over the past century since the expansion of canal irrigation. Groundwater levels and quality conditions vary across the plains during the irrigation and monsoon seasons (Qureshi, Shah, and Akhtar 2003). The Government's early strategy of constructing public SCARP (Pakistan's Salinity Control and Reclamation Projects) tubewells to manage waterlogging has been rapidly overtaken by an estimated 1 million unregulated private tubewells constructed for irrigation purposes. Some 87 percent of these tubewells run on diesel fuel, rather than unreliable and less flexible electricity supplies. When tubewells tap into brackish groundwater, they accelerate the secondary salinization of irrigated soils, which injures crops and reduces yields.

Food self-supply is an escalating concern in Pakistan. Food security can be defined in terms of the availability, access, and utilization of food supplies.[1] Although agricultural production and yields continue to grow, the annual population growth rate also remains high, at 2.2 percent. Per capita food supply varies from year to year (figure 1.2) and is below the global average of 2,797 kcal/capita/day. Despite increased food production, there has been no change over the past two decades in the estimated 25 percent of the population who are undernourished (FAOSTAT 2012). In 2004, the World Food Programme and the Sustainable Development Policy Institute prepared a national assessment of Food Insecurity in Rural Pakistan 2003 (WFP and SDPI 2004). The report concluded that (1) the common view that Pakistan's gross production could satisfy aggregate food needs belies a condition in which

Figure 1.2 Pakistan per Capita Food Supply, 1961–2009

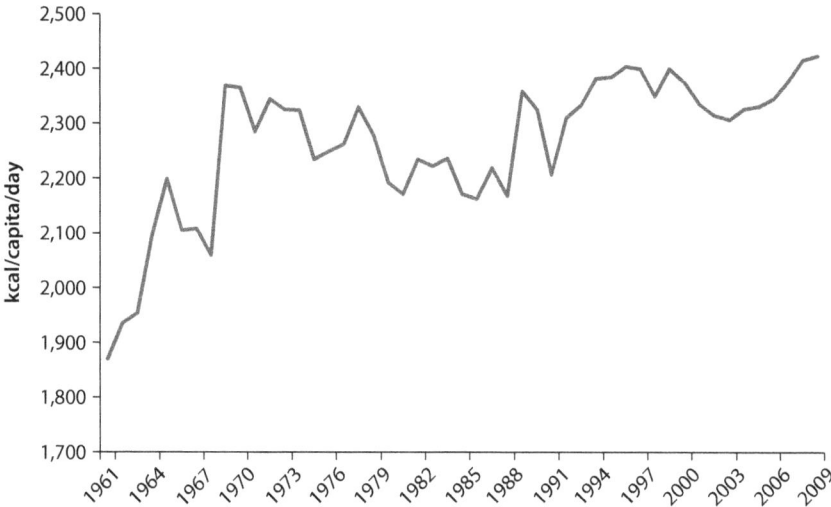

Source: FAOSTAT 2012.
Note: kcal = kilocalorie.

80 percent of the rural population experiences some level of food insecurity, and (2) the provinces vary significantly in the proportion of their districts that are food insecure, from a low of 29 percent in Punjab, to 65 percent in Sindh, and 85 percent in Balochistan. The National Nutrition Survey of 2011 (Bhutta 2012) reports that 57 percent of the population is food insecure. This report raises concerns about adverse childhood and lifelong developmental impacts from vitamin and micronutrient deficiencies.

Water and agricultural production depend on managing these many forms of resource variability and uncertainty. The overarching pattern that can be seen is that while the Indus Basin is richly endowed with land and water resources vital for the agricultural economy, it faces high levels of variability and uncertainty in climate, hydrology, agricultural sustainability, food consumption, and natural hazards.

Difficult Years for the Indus Basin: 2009–11

Each year the Indus Basin experiences a unique combination of weather, water, and agro-economic events. In 2009, the global economy and low-income people worldwide struggled to cope with the dramatic food price increases of 2008. Figure 1.3 indicates that the sharpest increases hit the staple food crops of wheat and rice, with wheat prices more than doubling in a year. Rice prices increased 60 percent between 2008 and 2009, after having already been increasing through the decade. Prices for high-value milk and meat products increased 24 percent. Nonfood crops like cotton increased by over 40 percent. Sugarcane has a lower base price, but it too increased by 24 percent that year. The causes of these shocks are debated as are future food price

Figure 1.3 Agriculture Prices, 2000–09

constant PRs per ton

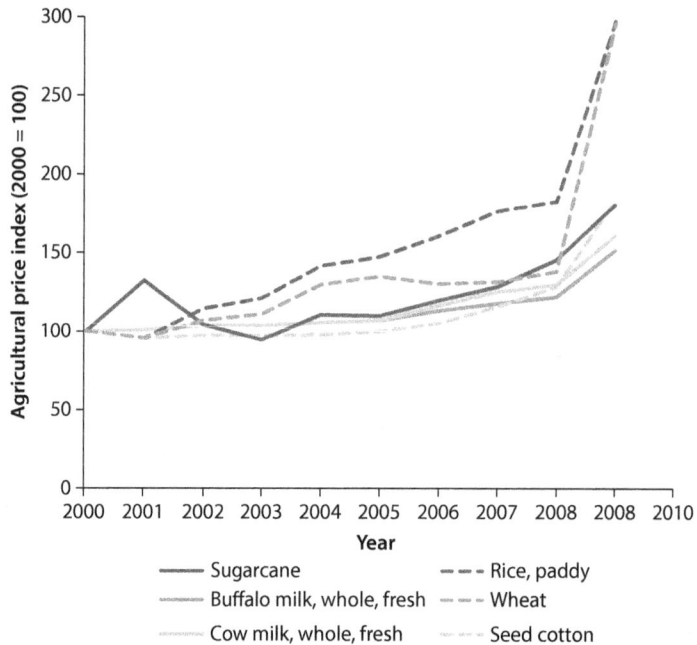

Legend:
- Sugarcane
- Buffalo milk, whole, fresh
- Cow milk, whole, fresh
- Rice, paddy
- Wheat
- Seed cotton

Source: FAOSTAT 2012.

projections. Some of the causes examined include increasing energy prices, biofuels policies, shifts toward more resource-intensive food consumption, reduced food stockpiles, and market distortions. While some global food prices dropped in 2009, they rose again in 2011. Following these events, the Government of Pakistan (GPPC 2009) issued a Task Force on Food Security report in 2009. The food security task force recommended policies to increase agricultural growth to at least 4 percent per year, coupled with pro-poor food and employment programs.

A weak monsoon hampered agricultural production in 2009. Average monsoon rainfall was about 30 percent below normal (PMD 2009). Drought was an extensive problem throughout the country. Punjab and Balochistan experienced net annual rainfall deficits of 26 and 41 percent, respectively. Sindh received around 50 percent less than average rainfall in August and September. While these deficits would normally have been offset by inflows from the Upper Indus and its tributaries, melt waters that year were also 15–30 percent below normal. These water constraints delayed winter wheat sowing until December 2009, posing risks to that staple food crop. At that time, diminished irrigation supplies led to questions about potential impacts of climate change and the associated concerns about the future of the glaciers in the Upper Indus. Increasing transboundary conflict over water development on the Jhelum and Chenab rivers exacerbated these concerns. Pakistan's increasing vulnerability to water scarcity was also highlighted in the literature (for example, Archer et al. 2010; Immerzeel, van Beek, and Bierkens 2010; Laghari, Vanham, and Rauch 2011). Around that time, the Government of Pakistan also issued a report of the *Task Force on Climate Change* (GPPC 2010).

In January 2010, a large landslide near the village of Attabad dammed the Hunza River valley, a tributary of the Upper Indus, inundating villages and destroying 19 km of the Karakoram Highway and cutting off the upper basin that produces seed potatoes as a cash crop from its markets down-country. Relief for this disaster included relocation of villagers and evacuation camps for those with irrigated lands downstream of the landslide. But these resettlement and reconstruction efforts were eclipsed by devastating floods later in the year.

The Indus River System Authority (IRSA), which is responsible for administering provincial water allocations under the 1991 Indus Water Accord, faced increasing conflicts over reservoir releases, 10-daily water allocations, and requests for canal closure, particularly between Punjab and Sindh. In 2011, there were increasing demands for releases for electricity generation, as well as objections to such releases. IRSA has had particular difficulty allocating water during periods of low inflows because of the structure of the Accord, which limited reservoir storage, water measurement constraints, and organizational capacity.

As late as June 2010, the Pakistan Meteorological Department (PMD) forecast a "normal (+10 percent)" monsoon. In late July, however, heavy rains fell over the Upper Indus main stem and the adjoining tributaries in the Kabul basin, causing extensive flash flooding in Khyber-Paktunkhwa province that cascaded through the districts that line the Indus from Punjab to Sindh and parts of Balochistan over the following month. Extremely high floods were recorded at the Chasma and Taunsa barrages, and a near historical flood peak was recorded at the Kotri barrage. Main stem levees were breached in many places, destroying the spring-season *kharif* crops of rice and cotton, as well as grain stores and seed for the winter-season *rabi* wheat planting. Additionally, flash floods and landslides triggered by the rain caused severe damage to infrastructure in the affected areas. More than 20 million people were adversely affected, with more than 1,980 dead and 2,946 injured. About 1.6 million homes were destroyed, and thousands of acres of crops and agricultural lands were damaged, some areas experiencing major soil erosion.

Massive international assistance was mobilized in response. A joint Asian Development Bank and World Bank (ADB and World Bank 2010) *Flood Damage and Needs Assessment* estimated that the total direct damages and indirect losses amounted to about US$10 billion; the agriculture, livestock, and fisheries sectors suffered the highest damages, calculated at US$5.0 billion.

As the 2011 monsoon season approached, the PMD forecast a slightly below normal (–10 percent) monsoon, with some areas expected to experience slightly above normal rainfall (+10 percent) (PMD 2011). However, heavy rains flooded the lower Indus Basin districts in Sindh and Balochistan, adversely affecting 5 million people, damaging 800,000 homes, and destroying 70 percent of the crops on flooded lands in what were already the most food insecure provinces in Pakistan (UNOCHA 2011). Although very different in hydroclimatic terms, the two floods of 2010 and 2011 had compounding damages on agricultural livelihoods and food security in the lower Indus Basin.

The years from 2009 through 2011 offer a perspective on the current challenges of water and food security, along with mounting future uncertainties that the federal and provincial governments must face. The prospects of climate change amplify these concerns. With growing populations and increasing water demand across all sectors, these risks must be anticipated and managed. This study will present a modeling framework for these purposes.

Literature Review on Indus Basin Modeling

This study follows a long legacy of research and planning for Pakistan's Indus Basin. The first major application of a multi-objective planning model for the Indus Basin was the World Bank's *Indus Special Study* of 1964–68, published as the three-volume report on *Water and Power Resources of West Pakistan: A Study in Sector Planning* (Lieftinck, Sadove, and Creyke 1968). It was an early use of linear programming and optimization modeling to weigh investment alternatives, which included Tarbela Dam and irrigation and agricultural development projects. The study developed a linear programming model to maximize the net economic benefits of production activities and projects in 54 canal commands under five different water budget conditions. Later, Duloy and O'Mara (1984) would develop the first version of the Indus Basin Model (IBM). It included farm production functions for different cropping technologies in the canal command areas and was based on a detailed rural household survey conducted in 1978. The analysis also linked hydrologic inflows and routing with irrigation systems, thereby showing where efficiencies could be gained in water allocation. Efficient allocation and economic pricing were shown to have substantial economic benefits that could support widespread tractor and tubewell investment, as well as increased farm income. Interestingly, the report concluded that by 1995, "all water resources [would be] fully utilized and thereafter gains would have to come from technical progress or substitution of more valuable crops in cropping patterns" (Duloy and O'Mara 1984, v).

This more streamlined version of the IBMR model ("R" was added for revised) was used in the Water and Power Development Authority's (WAPDA) next major basin analysis, known as the *Water Sector Investment Planning Study* (WSIPS) in the late 1980s, which focused on mid-term (10 year) development alternatives (WAPDA 1990). That study drew upon a 1988 farm survey to update farm production technologies and functions by canal command and nine agro-economic zones in the IBMR. The WSIPS evaluated a range of investment portfolios: no change, minimum investment, a Basic Plan of PRs 75 billion that optimized net economic benefits subject to a capital constraint, and a maximum plan contingent on additional investment funds being made available. A "plan generator" was also developed using mixed-integer programming techniques to assist in project scheduling and to ensure adherence to financial and other

Table 1.1 Increased Agricultural Production with and without the Basic Plan

Crop	Requirement 2000 (tonnes, thousands)	Without basic plan (% increase)	With basic plan (% increase)
Wheat	20,399	79	92
Rice	5,777	65	80
Sugarcane	47,204	80	98
Cotton	2,075	84	150
Pulses	991	83	92
Oilseeds	1,880	36	39

Source: WAPDA 1990.

macro-economic constraints (Ahmad and Kutcher 1992). The IBMR modeling showed what proportion of production targets for 1999–2000 could be met with and without the basic plan (table 1.1).

A detailed guide to the IBMR was written by Ahmad, Brooke, and Kutcher (1990). Ahmad and Kutcher (1992) followed this with a study looking at environmental considerations for irrigation planning, which incorporated salinity and groundwater variables in the IBMR water budget, flow routing, and management alternatives. This study noted slowing growth, increasing water scarcity, deteriorating infrastructure, extensive waterlogging and salinity, reduced growth of yields, and the high cost of drainage. It created large-scale groundwater and salt balance models and evaluated irrigation and drainage alternatives for achieving groundwater balance. The IBMR was later used for various projects and programs, for example, Kalabagh Dam (Ahmad, Kutcher, and Meeraus 1986); waterlogging and salinity under different scenarios of crop yield and tubewell investment in Sindh province (Rehman and Rehman 1993); and salinity management alternatives for the Rechna Doab region of Punjab (Rehman et al. 1997).

At about the same time, a team from WAPDA and the USEPA (United States Environmental Protection Agency) used the IBMR model to assess complex river basin management for Pakistan, which jointly analyzed general circulation model (GCM) climate scenarios along with WAPDA development alternatives (Wescoat and Leichenko 1992). The WAPDA-USEPA study of the Indus Basin examined temperature warming scenarios that ranged from an arbitrary +2°C to GCM-driven scenarios as high as +4.7°C. As precipitation was more uncertain and remains so, arbitrary ±20 percent scenarios were included. Upper basin snowmelt was modeled on the Jhelum River to generate inflows to the rim stations of the main IBIS. GCM warming scenarios in the upper Jhelum model simulated increased and earlier runoff. Water development scenarios were based on government plans for medium-term development that included the following scenarios: no projects, minimum development, and maximum development. The model was run with two different water allocation rules: 100 percent of historical water allocations and 80 percent of historical allocations with the remainder redirected to economically optimal uses. The net economic effects of these climate and water allocation scenarios are presented in table 1.2. All but

Table 1.2 Indus Basin Case Study Results: Total Economic Value-Added
PRs, billions

Case study scenarios	No climate change	+2°C 0% P	+2°C +20% P	+2°C −20% P	GISS[a] +30% P	GFDL[b] +20% P
1988 water management						
100% allocation	90.515	88.114	92.797	Infeasible	89.475	88.643
80% allocation	94.203	n.a.	n.a.	88.829	n.a.	n.a.
2000 with no new projects						
100% allocation	Infeasible	Infeasible	136.923	Infeasible	Infeasible	132.903
80% allocation	138.641	134.862	n.a.	127.647	136.164	n.a.
2000 with minimum investment						
100% allocation	136.511	Infeasible	140.184	Infeasible	Infeasible	134.854
80% allocation	143.162	139.417	n.a.	133.882	138.956	n.a.
2000 with maximum investment						
100% allocation	143.434	Infeasible	147.178	Infeasible	Infeasible	138.996
80% allocation	149.202	146.351	n.a.	140.593	144.585	n.a.

Source: Wescoat and Leichenko 1992.
Note: n.a. = not applicable.
a. NASA (U.S. National Aeronautics and Space Administration) Goddard Institute for Space Studies.
b. NOAA (U.S. National Oceanic and Atmospheric Administration) Geophysical Fluid Dynamics Laboratory.

the "+2°C +20% P" scenario had a negative impact on the objective function. Impacts ranged from −7.9 to +2.7 percent of total value-added (or from about one to three years of economic growth at 3 percent). Changing the allocation rule had a greater positive economic effect of +4.0 to +4.9 percent. These gains were largely eliminated by climate change scenarios.

This early study was also able to compare the potential economic impacts of climate change scenarios on different investment portfolios. For example, it showed that climate change diminished the net economic benefits of the minimum investment plan from 40 to 100 percent. This earlier work also demonstrated that, with some exceptions, the Indus Basin irrigation baseline seemed relatively robust in the face of the types of climate variability considered. This may reflect high levels of inflow and monsoon variability, system redundancy, groundwater availability, and/or compensating farming decisions in the optimization model.

Habib (2004) used the HYDRAM model to scope out the reallocation opportunities in the Indus Basin. This study built on a detailed analysis of water budgets and canal diversions (Kaleemuddin, Habib, and Muhammad 2001; Tahir and Habib 2001). The study identified important network and operational constraints, flexibility, and tradeoffs for meeting water allocation and delivery targets. Khan et al. (n.d.) used the Soil and Water Assessment Tool (SWAT) model in a regional watershed analysis of the Upper Indus Basin. This study prepared a digital elevation model of the watershed, along with large-scale land use and soil maps to model agricultural hydrology in the Upper Indus. Also, a 2002 version of the IBMR was used to assess economic and water management benefits of raising Mangla Dam by different heights (Alam and Olsthoorn 2011). Finally, the Global Change Impact Study Centre (GCISC) in Pakistan undertook a number of adaptation studies (for example, Ali, Hasson, and Khan 2009). Using a sophisticated crop model, these studies focus primarily on examining how climate change may impact wheat and rice yields and production (Iqbal et al. 2009a, 2009b, 2000c).

Based on this literature review, the following four needs stand out: (1) a wider perspective on the policy environment, (2) expansion of the scientific basis for snow and ice hydrology in the upper basin, (3) advanced and updated modeling of hydroclimatic impacts on water and food systems using the IBMR, and (4) agro-economic modeling with a more sophisticated computable general equilibrium (CGE) and social accounting matrix (SAM) approach. A framework for addressing these gaps will be described here and in later chapters.

Study Approach: A Framework for Integrated Water and Agriculture Assessment

The objective of this study is to assess the potential impacts of climate risks and various alternatives for minimizing those impacts on water and food security in the Indus Basin of Pakistan. The study analyzes interrelationships among the climate, water, and agriculture sectors to gain a better understanding of how these factors are linked in order to help guide the prioritization and planning of future investments in these sectors. Attention is also given to analysis by province, as provinces are the primary level of water and agricultural governance in the federal system. Analytically, the study objective is achieved by integrating several different modeling environments: a model of Upper Indus snow and ice hydrology, an agro-economic optimization model of the IBIS, and an updated CGE model of Pakistan's wider macro-economy. This integration of models helps frame the recommendations for strengthening water, climate, and food security planning, policies, and research priorities for the Indus Basin. The five key tasks for this analysis are shown in figure 1.4.

First, this study will review the major challenges and the current water and agriculture context, plans, and policies. Chapter 2 surveys the current policy environment for addressing water and agricultural issues in a changing climate. This policy environment is shaped by economic development plans at the national and provincial levels, sector plans for water and agriculture (from long-term, multi-decade plans to medium-term and annual plans), and recent cross-cutting policy documents on climate change. This policy review establishes the context for scientific and modeling efforts in subsequent chapters.

Second, the study will assess glacier-melt and snowmelt dynamics in the upper Indus Basin and implications for downstream inflows. Chapter 3 examines the state of the science associated with the snow and ice hydrology in the Upper Indus Basin and reviews the literature and data available on the present and projected role of glaciers, snow fields, and stream flow. A simple hydrologic model is developed to estimate the relative contributions of glaciers and snow to

Figure 1.4 Framework for Integrated Water and Agriculture Assessment

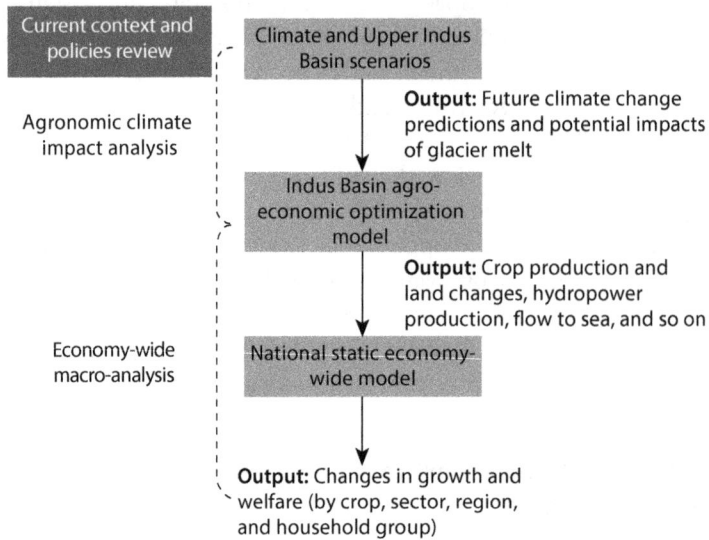

| Current context and policies review |
| Climate and Upper Indus Basin scenarios |

Output: Future climate change predictions and potential impacts of glacier melt

Agronomic climate impact analysis

| Indus Basin agro-economic optimization model |

Output: Crop production and land changes, hydropower production, flow to sea, and so on

| National static economy-wide model |

Economy-wide macro-analysis

Output: Changes in growth and welfare (by crop, sector, region, and household group)

the hydrologic regimes of the Upper Indus Basin. Topographic imagery is used to delineate basin areas, area versus altitude relationships, hypsometry, and ablation processes. The results of these analyses contribute to climate scenario construction for the downstream IBIS modeling.

Third, climate scenarios are constructed for analysis with the Indus Basin Model Revised (IBMR). Chapter 4 examines the literature and available data on hydroclimatic variability and change on the Indus Basin plains. It compares historical fluctuations in climatic and hydrologic variables in the Indus Basin. Scenarios of climate change derived from GCMs are also reviewed, including the generation of future scenarios of changing snow and ice melt in the Upper Indus Basin.

Fourth, two primary models are described in chapter 5. The IBMR model is a powerful agro-economic optimization model used and refined over three decades by the World Bank and Government of Pakistan. The first part of chapter 5 describes the model, the updates made for this study, and the dependent variables in the model output. Sensitivity of the model to key water, agricultural, and land use indicators is also given. The second part of chapter 5 describes the use of an updated social accounting matrix and CGE model to explore the economy-wide impacts of changes in the agriculture sector. This model includes 49 economic activities and 48 commodities. It includes the quantities and prices of agricultural inputs and agricultural industries beyond crop production, which offers a more complete assessment of economic impacts. The model also differentiates across 19 types of households by farm size, tenancy, and poverty level to give more detailed insights into social impacts.

Fifth, the results of the various scenarios using these models and policy and investment implications are discussed in chapter 6. Chapter 7, the final chapter, draws together the findings from the chain of analyses. It distinguishes between the relative significance of different scenarios, impacts, and adaptations, and highlights recommendations for research, planning, and policies that can help to expand the range of options for Indus Basin management.

Note

1. *Food availability* is defined as having sufficient quantities of food on a consistent basis. *Food access* is defined as having sufficient resources to obtain appropriate foods for a nutritious diet. *Food use* is defined as appropriately using food for one's basic nutrition and care, as well as having adequate water and sanitation (FAO World Food Summit 1996).

References

ADB (Asian Development Bank) and World Bank. 2010. "Pakistan Floods 2010 Damage and Needs Assessment." Paper presented at the Pakistan Development Forum, Islamabad, November 14–15.

Ahmad, M., A. Brooke, and G. P. Kutcher. 1990. *Guide to the Indus Basin Model Revised.* Washington, DC: World Bank.

Ahmad, M., and G. P. Kutcher. 1992. "Irrigation Planning with Environmental Considerations: A Case Study of Pakistan's Indus Basin." World Bank Technical Paper 166, World Bank, Washington, DC.

Ahmad, M., G. Kutcher, and A. Meeraus. 1986. *The Agricultural Impact of the Kalabagh Dam (As Simulated by the Indus Basin Model Revised).* Vols. I and II. Washington, DC: World Bank.

Alam, N., and T. N. Olsthoorn. 2011. "Sustainable Conjunctive Use of Surface and Groundwater: Modeling on the Basin Scale." *International Journal of Natural Resources and Marine Sciences* 1: 1–12.

Ali, G., S. Hasson, and A. M. Khan. 2009. *Climate Change: Implications and Adaptation of Water Resources in Pakistan.* Research Report GCISC-RR-13, Global Change Impact Study Centre, Islamabad.

Archer, D. R., N. Forsythe, H. J. Fowler, and S. M. Shah. 2010. "Sustainability of Water Resources Management in the Indus Basin under Changing Climatic and Socio Economic Conditions." *Hydrology and Earth System Sciences* 14: 1669–80.

Bhutta, Z. 2012. *Pakistan—National Nutrition Survey 2011.* Karachi, Pakistan: Aga Khan University Pakistan Medical Research Council Nutrition Wing, and Ministry of Health.

Brakenridge, G. R. 2012. "Global Active Archive of Large Flood Events." Dartmouth Flood Observatory, University of Colorado, Boulder, CO (accessed January 25, 2013). http://floodobservatory.colorado.edu/Archives/index.html.

COMSATS (Commission on Science and Technology for Sustainable Development in the South). 2003. *Water Resources in the South: Present Scenario and Future Prospects.* Islamabad: COMSATS.

Duloy, J. H., and G. T. O'Mara. 1984. "Issues of Efficiency and Interdependence in Water Resource Investments: Lessons from the Indus Basin of Pakistan." World Bank Staff Working Paper 665, World Bank, Washington, DC.

FAO World Food Summit. 1996. *Rome Declaration on World Food Security.* Food and Agriculture Organization of the United Nations, Rome, Italy.

FAOSTAT (Food and Agricultural Organization of the United Nations Statistical Database). 2012. Database of Food and Agriculture Organization of the United Nations, Rome. http://faostat.fao.org.

Giosan, L., P. D. Clift, M. G. Macklin, D. Q. Fuller, S. Constantinescu, J. A. Durcan, T. Stevens, G. A. T. Duller, A. R. Tabrez, K. Gangal, R. Adhikari, A. Alizai, F. Filip, S. VanLaningham, and J. P. M. Syvitski. 2012. "Fluvial Landscapes of the Harappan Civilization." In *Proceedings of the National Academy of Sciences.* doi: 10.1073/pnas.1112743109. http://www.pnas.org/content/109/26/E1688 (accessed January 28, 2013).

GPPC (Government of Pakistan, Planning Commission). 2009. *Final Report of the Task Force on Food Security.* Islamabad.

———. 2010. *Task Force on Climate Change Final Report.* Islamabad.

Habib, Z. 2004. "Scope for Reallocation of River Waters for Agriculture in the Indus Basin." PhD thesis. Ecole Nationale du Genie Rural, des Eaux et des Forets, Paris.

Immerzeel, W. W., L. P. H. van Beek, and M. F. P. Bierkens. 2010. "Climate Change Will Affect the Asian Water Towers." *Science* 328 (5984): 1382–85.

Iqbal, M. M., M. A. Goheer, S. A. Noor, H. Sultana, K. M. Salik, and A. M. Khan. 2009a. *Climate Change and Rice Production in Pakistan: Calibration, Validation and Application of CERES-Rice Model.* Research Report GCISC-RR-15, Global Change Impact Studies Centre (GCISC), Islamabad.

———. 2009b. *Climate Change and Agriculture in Pakistan: Adaptation Strategies to Cope with Negative Impacts.* Research Report GCISC-RR-16, Global Change Impact Studies Centre, Islamabad.

———. 2009c. *Climate Change and Wheat Production in Pakistan: Calibration, Validation and Application of CERES-Wheat Model.* Research Report GCISC-RR-14, Global Change Impact Studies Centre, Islamabad.

Kaleemuddin, M., Z. Habib, and S. Muhammad. 2001. "Spatial Distribution of Reference and Potential Evapotranspiration." Working Paper 24, Pakistan Country Series number 8, International Water Management Institute, Lahore, Pakistan.

Khan, A. D., J. G. Arnold, M. DiLuzio, and R. Srinavasan. n.d. "GIS Based Hydrologic Modeling of Upper Indus Basin." Unpublished manuscript.

Laghari, A. N., D. Vanham, and W. Rauch. 2011. "The Indus Basin in the Framework of Current and Future Water Resources Management." *Hydrology and Earth System Sciences* 14 (8): 2263–88.

Lieftinck, P., R. A. Sadove, and T. A. Creyke. 1968. *Water and Power Resources of West Pakistan: A Study in Sector Planning.* Baltimore, MD: Johns Hopkins Press.

Mughal, M. R. 1997. *Ancient Cholistan: Architecture and Archaeology.* Lahore, Pakistan: Ferozsons.

PMD (Pakistan Meteorological Department). 2009. *Flood Report 2009.* Flood Forecasting Division, Pakistan Meteorological Department, Lahore, Pakistan (accessed May 1, 2012). http://www.pakmet.com.pk/FFD/cp/fr2009.pdf.

———. 2011. "Outlook for Monsoon Season (July–September 2011)." National Weather Forecasting Centre, Pakistan Meteorological Department, Lahore, Pakistan. http://pakmet.com.pk/MON&TC/Monsoon/monsoon2010.html.

Qureshi, A. S., T. Shah, and M. Akhtar. 2003. "The Groundwater Economy of Pakistan." Working Paper 64, International Water Management Institute, Lahore, Pakistan.

Rehman, A., and G. Rehman. 1993. *Strategy for Resource Allocations and Management across the Hydrologic Divides*. Volume 3 of *Waterlogging and Salinity Management in the Sindh Province, Pakistan*. Report R-T0.3, International Irrigation Management Institute, Lahore, Pakistan.

Rehman, G., M. Aslam, W. A. Jehangir, A. Rehman, A. Hussain, N. Ali, and H. Z. Munawwar. 1997. *Salinity Management Alternatives for the Rechna Doab, Punjab, Pakistan*. Volume 3 of *Development of Procedural and Analytical Links*. Report R-21.3, International Irrigation Management Institute, Lahore, Pakistan.

Tahir, Z., and Z. Habib. 2001. "Land and Water Productivity: Trends across Punjab Canals." IWMI Working Paper 14, Pakistan International Water Management Institute, Lahore, Pakistan.

UNOCHA (United Nations Office for the Coordination of Humanitarian Affairs). 2011. *Pakistan Monsoon 2011*. Situation Report 14, Islamabad (accessed May 1, 2012). http://pakresponse.info/LinkClick.aspx?fileticket=nu4xv8K2MZ4%3D&tabid=87&mid=539.

WAPDA (Water and Power Development Authority). 1990. *Water Sector Investment Planning Study (WSIPS)*. 5 vols. Lahore, Pakistan: Government of Pakistan Water and Power Development Authority, Lahore.

Wescoat, J., and R. Leichenko. 1992. "Complex River Basin Management in a Changing Global Climate: Indus River Basin Case Study in Pakistan—A National Modeling Assessment." Collaborative Paper 5, Center for Advanced Decision Support for Water and Environmental Systems, University of Colorado, Civil, Environmental, and Architectural Engineering, Boulder, CO.

WFP and SDPI (World Food Programme and the Sustainable Development Policy Institute). 2004. *Food Insecurity in Rural Pakistan 2003*. Islamabad: World Food Program VAM Unit.

World Bank. 2012. "World Development Indicators Databank (WDI)." http://databank.worldbank.org/ddp/home.do?Step=12&id=4&CNO=2.

Wright, R. 2010. *The Ancient Indus: Urbanism, Economy and Society*. Cambridge, U.K.: Cambridge University Press.

CHAPTER 2

The Current Water and Agriculture Context, Challenges, and Policies

Key Messages

- Multiyear storage in the Indus Basin remains limited.
- Water and food demands are likely to increase on a per capita basis and in aggregate terms, as population increases. Reliance on groundwater resources will continue. Falling water tables and increased salinity in many places may worsen.
- An array of allocation entitlements economically constrains the waters available for agricultural production and coping with climatic risks.
- Low water-use efficiencies and agriculture productivities are top concerns.
- A common set of water and agricultural policy challenges is complicated by several dynamic stresses and institutional shifts, including constitutional devolution from national to provincial levels.
- Most national and provincial development plans continue to focus on the role of infrastructure in addressing challenges of water and food security.
- Recent policy documents highlight the increasing importance of improving irrigation efficiency, improvement of yields, and the socioeconomic distribution of development opportunities and benefits, including food security.
- The important role that water management plays in the productivity of the agriculture sector is recognized in many different forums and policy reports. However, these linkages are not always comprehensively addressed (with systems-based models) in federal and provincial planning documents and budgets.

The Indus Basin Irrigation System (IBIS) has undergone profound changes and experienced increasing stresses in recent years. Several recent studies have heightened awareness of Indus water resources issues, notably the World Bank study *Pakistan's Water Economy: Running Dry* (Briscoe and Qamar 2006).[1] That study convened a team of experts to identify broad challenges and strategic choices facing the water sector in Pakistan.

The key challenges that this modeling framework will examine in the context of climate risks are (1) limited water storage, (2) problematic trends in surface water and groundwater use, (3) inflexible and uncertain water allocation institutions, and (4) low water-use efficiencies and productivity. This chapter will also look at the various national policies and development plans to address these water and food security concerns.

Limited Water Storage

It is well known that South Asian countries have a lower proportion of water storage and hydropower development than other regions of the world, both in relation to their geographical potential for storage and power generation and in relation to per capita water and energy use (figure 2.1).

No major reservoir storage projects have been constructed since the completion of Tarbela Dam in 1976, and the system does not have multiyear carryover storage. The total storage is about 11 million acre-feet (MAF), representing about 10 percent of the total inflow in the system. This storage total is likely to decline with increased sedimentation into these reservoirs. This may constrain the quantity and timing of water releases for canal irrigation, and could have the greatest economic impact on the agricultural sector (Amir 2005a, 4). Moreover, figure 2.2 shows how storage per capita is likely to decline with continued population growth. Historically, reservoirs in Pakistan have been operated first for their irrigation benefits, and secondarily for their hydropower generation benefits. Interestingly, there appears to have been a major shift in sector benefit ratios. Work by Amir (2005b) suggests that the hydropower benefits of proposed dams at Basha and Kalabagh are estimated to be substantially greater than their irrigation benefits. The benefits from flood control are estimated to be even smaller. Amir (2005a) qualifies this generalization by noting that (1) Tarbela Dam has provided 22 percent of the surface irrigation deliveries in Punjab alone and has

Figure 2.1 Water Storage per Capita in Semi-Arid Countries

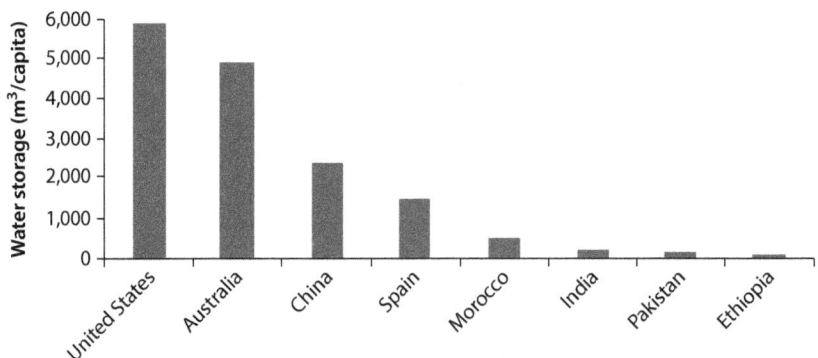

Source: Briscoe and Qamar 2006.

Figure 2.2 Water Storage per Capita over Time

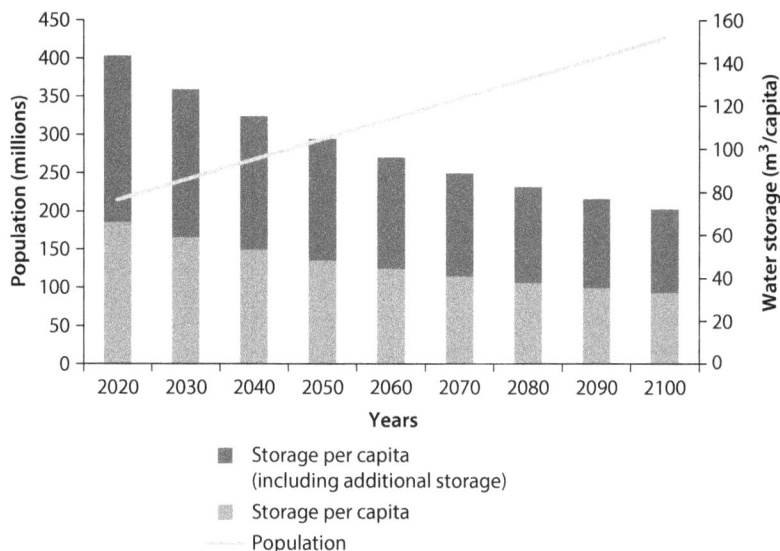

had substantially greater agricultural benefits than predicted, (2) reservoirs reduced the variability of water supplies for *rabi* crops and for delivering water to eastern canal commands during drought years, and (3) in wet years they have helped expand irrigated area and reduce the degree of deficit irrigation.

The *WAPDA 2025 Plan* (WAPDA 2004) identifies 22 storage projects in the IBIS. By 2011, some 800+ hydropower projects were identified, which would increase the nation's estimated power capacity by some 30 percent (Government of Pakistan Private Power and Infrastructure Board 2011; Siddiqi et al. 2012). This analysis is particularly relevant in the wider South Asian regional context where there is now a "race to the top" to develop new reservoirs throughout the Hindu Kush-Himalayan region. Although annual flood control benefits are estimated to be far smaller than those of irrigation or power, they can be periodically significant, as evidenced by the economic impact of the floods of 2010. ADB and World Bank (2010) identifies the following key issues related to flood management: (1) the deferred maintenance of flood embankments resulting in structural failures, (2) insufficient storage capacity to absorb flood peaks, (3) lack of response mechanisms to early warnings, (4) need for expanding flood early warning systems, and (5) encroachment into flood plains and riverine areas. Post-flood assessments underscored the imperative of nonstructural as well as structural measures, and their relative and joint significance have yet to be established.

Problematic Trends in Surface Water and Groundwater Usage

A second area of concern is the changing relationship between surface and groundwater irrigation, with the underlying issues of (1) declining per capita water availability due to continuing population growth and (2) increasing rates

Figure 2.3 Pakistan Population Prospects under Low, Medium, and High Scenarios

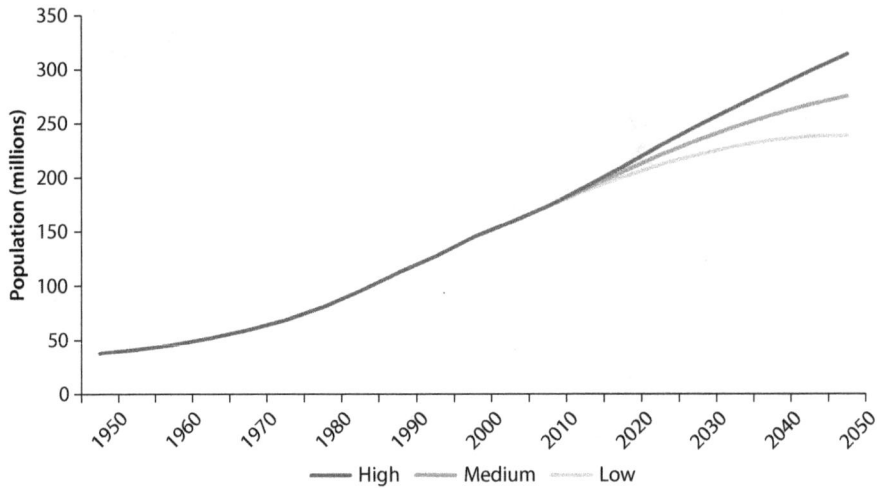

Source: UN Population Division 2012.

of groundwater pumping. Although the population growth rate in Pakistan has been declining, it is still 1.8 percent annually, which portends escalating demand for water, food, and fiber crops (World Bank 2012b). Even with the relative decline in the population growth rate, today's 174 million are projected to be 238 million to 314 million by 2050 (UN Population Division 2012) (figure 2.3). Will land and water resources suffice for this population? Moreover, water availability per capita has fallen drastically from 5,650 m^3 in 1951 to 1,000 m^3 in 2010. And by 2025, this number is projected to fall to 800 cubic meters (m^3) per capita (GPPC 2007), well below the 1,000 m^3 per capita limit below which the supply is defined as "water insecure" (Falkenmark et al. 2007).

Land use from the late-19th to early 20th century involved a dramatic shift from a pastoral landscape punctuated by localized shallow well irrigation to the largest contiguously managed canal irrigation system in the world (Bedi 2003). By the second half of the 20th century, private tubewell development had accelerated (figure 2.4) as a means to reduce waterlogging and provide a more reliable and timely water supply for irrigation (Michel 1967). These processes enabled the reclamation of agricultural land to grow. Cropping intensity also increased in many areas from one to two crops per year, which has contributed to the continuing growth of withdrawals for agriculture. More recent data suggest that tubewell development has been leveling off during the current decade, perhaps due to increasing pumping costs, unreliable fuel supplies (mainly diesel), and decreasing groundwater quality.

Irrigated land increased at an average annual rate of almost 1 percent from 1992 to 2008 (Government of Pakistan, Ministry of Food, Agriculture, and Livestock 2010). Figure 2.5 shows the change in irrigation source over time (Van Steenbergen and Gohar 2005). Indeed, most farmers are currently using

Figure 2.4 Growth in Use of Tubewells, 1960–2003

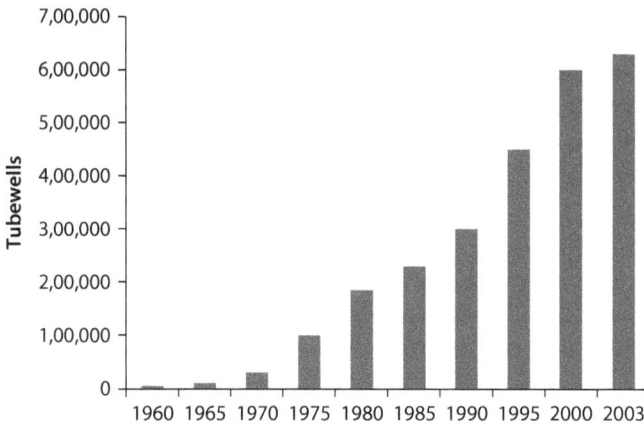

Sources: Van Steenbergen and Gohar 2005; in Briscoe and Qamar 2006.

Figure 2.5 Growing Role of Groundwater Irrigation, 1960–99

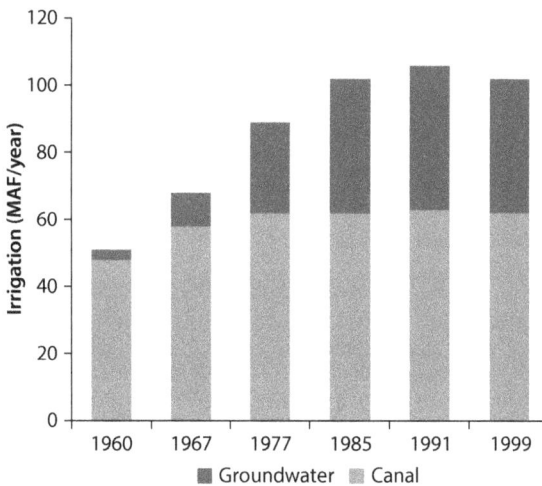

Sources: Van Steenbergen and Gohar 2005; in Briscoe and Qamar 2006, 41.
Note: MAF = million acre-feet.

a combination of canal and tubewell water, while a smaller proportion relies solely on tubewells. Food and Agricultural Organization of the United Nations Statistical Database (FAOSTAT 2012) data estimate groundwater withdrawals at 34 percent of total withdrawals for all uses, which is probably conservative. Canal irrigation remains enormously important, though it has been slowly declining as the predominant source of water. Further opportunities for expanding irrigation into areas of rainfed (*barani*) cultivation are limited, although they include expansion of private tubewell irrigation, watercourse extension, and high-efficiency irrigation technologies that can operate on uneven terrain (for example, drip systems). These opportunities vary by province.

Increasing reliance on groundwater is no doubt related to problems of waterlogging, salinity, and, in some areas, groundwater depletion. Before an extensive canal irrigation network was developed in the Indus Basin, it was a land of monsoon-flooded riparian corridors between the dry upland plains of the great *doabs* and deserts. As irrigation historians of the Indus have shown, the colonial and post-colonial canal system had extensive seepage and spread vast quantities of water over the land that raised the groundwater table dramatically (figure 2.6; Gilmartin 1994).

Drawing down groundwater can improve the waterlogging situation, but it can also increase pumping costs or it can tap into increasingly brackish waters. Figure 2.7 shows the particularly high proportion of shallow groundwater in Sindh province, although it varies across years.

A closer look at recent well records of water table depths shows significant differences over space and time. In the northern district of Sialkot in Punjab, for example, where water table levels are relatively high, water tables have fluctuated between 4 and 16 feet below the surface (figure 2.8).

The overall regional pattern shows decreasing water tables, thus decreased waterlogging, in the basin. To understand the potential impacts on agricultural production and yields, the variability of water levels and waterlogging must be compared with changing irrigation patterns and climatic conditions.

The areas affected by waterlogging and salinity have been monitored, but the costs of this environmental degradation are difficult to estimate. A recent

Figure 2.6 Historical Changes in Groundwater Levels

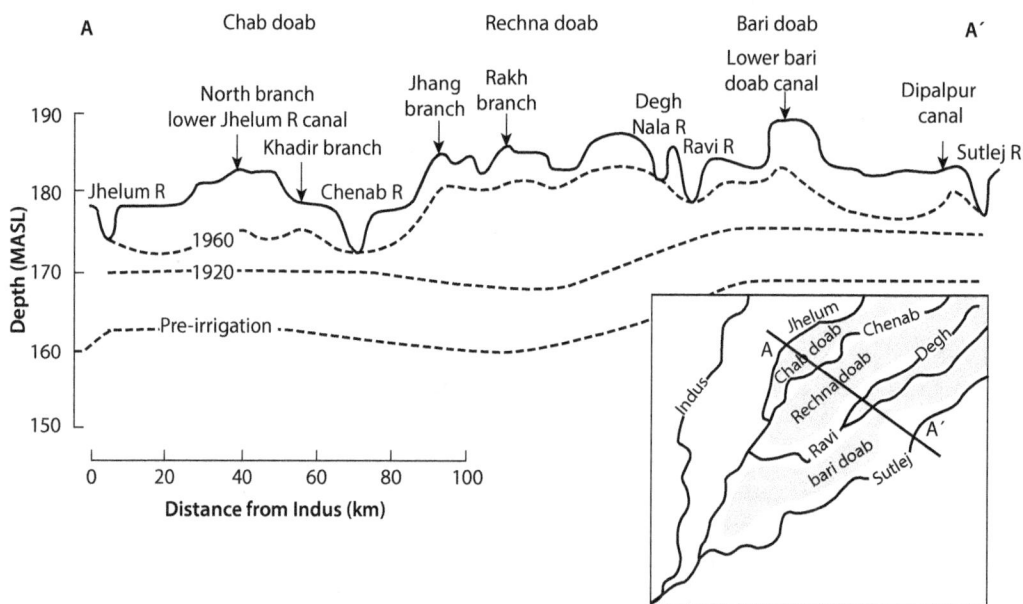

Sources: Bhutta and Smedema 2005; in Briscoe and Qamar 2006.
Note: MASL = meters above sea level, km = kilometers.

Figure 2.7 Depth of Water Table by Province

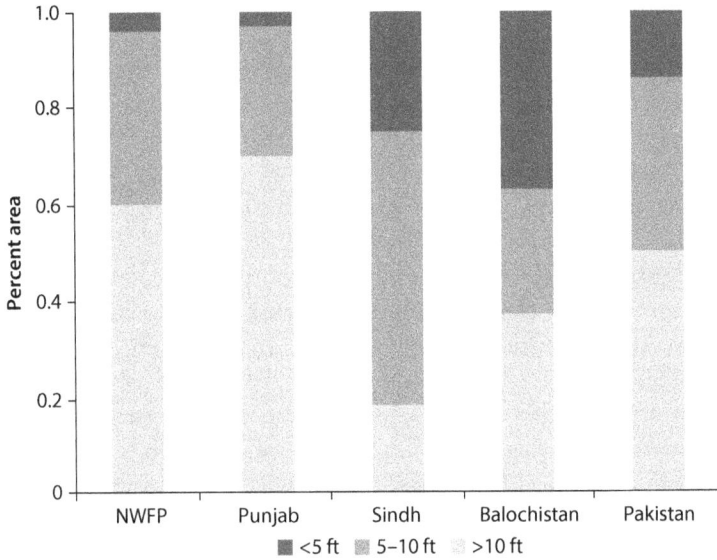

Source: Van Steenbergen and Gohar 2005; in Briscoe and Qamar 2006.
Note: NWFP = North-West Frontier Province.

Figure 2.8 Water Table Trends for Wells in Sialkot District of Punjab, 2003–08

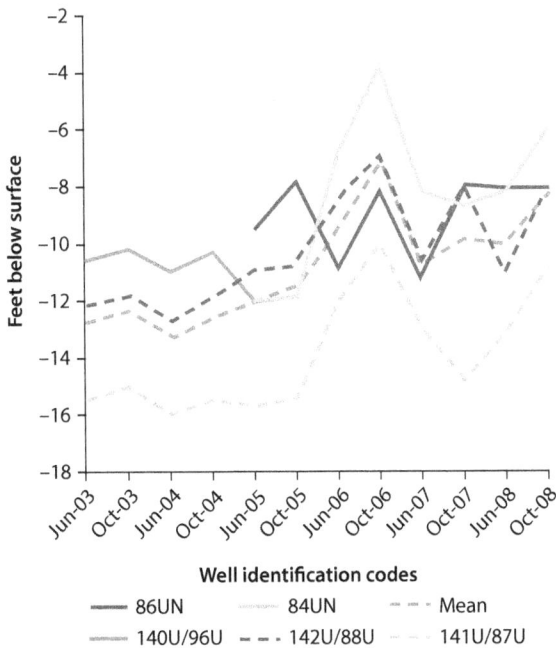

Source: Punjab Irrigation Department 2009, 12.

study by Bhutta and Smedema (2005) noted that the direct annual agricultural damage (not counting the lost opportunities of more profitable land use) is estimated to be on the order of PRs 20 billion per year. Waterlogging and salinity have also adversely affected public health and sanitary conditions in the villages. A more recent national estimate of the economic impacts of salinity on agricultural production examined two scenarios, one that emphasizes cotton planting on the most saline lands and the other, wheat. Estimated crop production losses ranged from PRs 30 to 80 billion in 2004 prices (World Bank 2006, 26).

Groundwater quality issues are even more spatially complex and pose variable threats to agricultural sustainability. Overall salt balance models have been estimated, but as Ahmad and Kutcher (1992) have shown, the main challenge is modeling the dynamic spatial distribution and transport of salts through the irrigation system. As with waterlogging hazards, salinity hazards tend to accumulate downstream, affecting as much as 50 percent of the land in Sindh. However, recent village surveys indicate improvements in waterlogging and salinity in upper Sindh, as contrasted with increasing concerns in the deltaic region of Badin and Thatta districts (Berger and IAC 2011). Even in Punjab, salinity conditions vary by sampling well locations within a district and over time.

Inflexible and Uncertain Water Allocation Institutions

An array of entitlements—from individual timeshare water rights (*warabandi*), to canal indents, a provincial water accord, and the Indus Waters Treaty (IWT)—shape the waters available for increasing agricultural production and coping with climatic risks. While private groundwater pumping is not regulated, surface water allocation institutions were designed with limited flexibility that constrain production under conditions of hydroclimatic variability and changing crop production technologies and functions. Additionally, these inflexible institutions have been routinely subverted to effect changes that privilege one group over another, undermine trust, and increase singular perceptions of even more widespread distortions. Two examples follow.

The canal water timeshare system is said to have been designed by colonial engineers to operate continuously under variable flow conditions with minimal involvement of the irrigators affected (Gilmartin 1994). Outlets (moghas) located on distributary and minor channels had fixed outlet sizes that were opened and closed for their respective shares based on farm size and duty of water under the 1873 Canal Act (figure 2.9). Over time, this system has been increasingly distorted through the modification of outlet sizes and timings to produce systematic inequities in waters that are over-allocated at the "head" of a canal leaving those at its "tail" deprived. Interestingly, Bhatia (2005) indicates that the presumed benefits of excessive diversions at the canal head may not always translate into net economic benefits at the canal head, as they can be offset both by excess water deliveries and less intensive or ineffective on-farm management. In addition, head-middle-tail inequities vary enormously in their magnitude and

Figure 2.9 Canal Water Distribution System

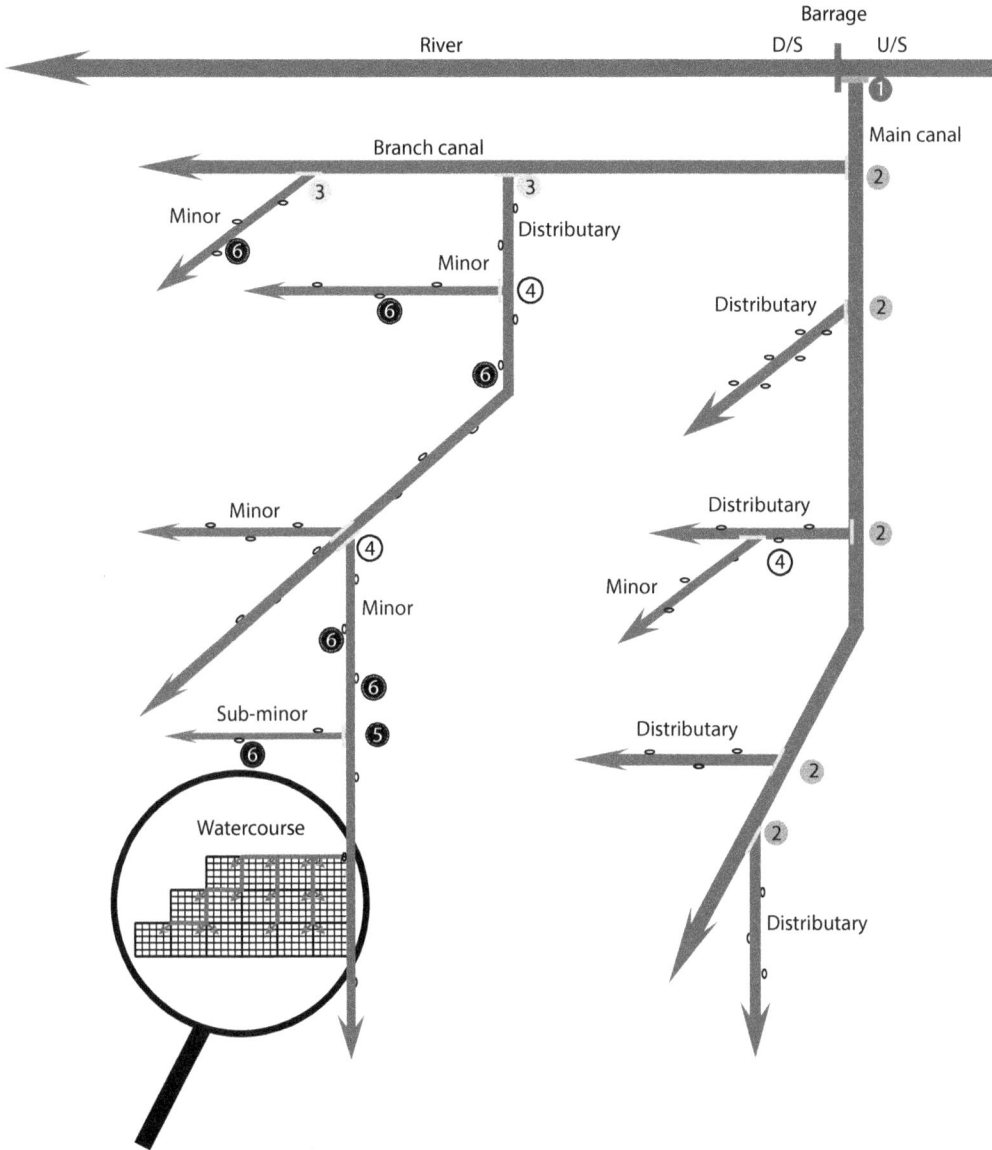

Barrage

River D/S U/S

Main canal

Branch canal

Distributary

Minor

Minor

Distributary

Distributary

Minor

Minor

Distributary

Sub-minor

Minor

Distributary

Watercourse

Distributary

Transaction point	Water transaction	
	From	To
❶	Barrage	Main canal
②	Main canal	Branch canals/ Distributaries
③	Branch canal	Distributaries/Minors
④	Distributary	Minors/Watercourses
❺	Minor	Sub-minors/Watercourses
❻	Sub-minor	Watercourses

———	Barrage
———	Head work
········	Head regulator
○	Mogha/Outlet
Field	Turnout

Source: Blackmore and Hasan 2005, 49.

impact on crop production (Hussain et al. 2003), variations that are sometimes interpreted as "low yields."

Indus Water Treaty

Upon national independence in 1947, east and west Punjab were partitioned, and former princely states such as Jammu and Kashmir were placed in transitional status, which cut across the headwaters of the Indus tributary headwaters and created uncertainties for basin development in both Pakistan and India. Eight years of intensive negotiation with support from the World Bank yielded the IWT[2] of 1960 and a bold engineering and investment framework for the Indus Basin Development Programme (IBDP) in Pakistan (see Michel 1967 for a detailed history). The treaty allocated upper basin flows of the "eastern rivers" to India (Beas, Sutlej, and Ravi) and historical upper basin flows of the "western rivers" to Pakistan (Chenab, Jhelum, and Indus), with detailed specifications on future upstream development of the western rivers.

The impact of the IWT and IBDP in reshaping the IBIS cannot be overstated. They enabled construction of replacement works in Pakistan that included Pakistan's two major storage dams (Mangla and Tarbela) and link-canals to transfer inflows from the western rivers to canal commands formerly supplied by the eastern rivers in Pakistan (see Wescoat, Halvorson, and Mustafa 2000 for a 50-year review of Indus Basin development). The Indus Basin Model used in this study was created to help guide investment in this highly complex agro-economic system.

The IWT (World Bank 2012a) has endured various stresses over time, and it has come under new pressures over upper basin development. In 2007, the IWT article that provides for the appointment of a neutral expert on issues that cannot be resolved by the parties was employed for the first time in the case of Baglhiar Dam on the Chenab River.[3] In 2010, an International Court of Arbitration was convened to address Pakistan's objections to the Kishanganga project under Article IX and Annexure G of the IWT. The IWT will likely continue to be tested as questions of climate risk, water, and food production become transboundary concerns. Although some question the future robustness of the IWT in light of the increasing scale of hydropower development and other trends, it has up till now worked reasonably as envisioned regarding international disputes.

Interestingly, multi-track efforts are under way among scientists, scholars, and former officials in India and Pakistan, the potential of which may be greater than has been realized to date.[4] Stochastic analysis of the joint and cumulative hydrologic and environmental effects of upper basin climate, runoff, and hydropower development processes could help identify paths for data exchange, confidence-building, data-driven negotiations, and expanding the range of choices among management alternatives.

Indus Water Accord

Since 1991, water inflows have been apportioned among the provinces by the Indus Water Accord. The Accord of 1991 allocated annual flows among the provinces based on a five-year record of pre-Accord historical canal diversions. The Accord, which was based on the *assumed* average flow of 114.35 MAF of water in the Indus system, allocated 55.94 MAF of water to Punjab and 48.76 MAF to Sindh province, the remaining 9.65 MAF was divided between North-West Frontier Province (NWFP, currently known as Khyber Pakhtunkhwa) and Balochistan provinces (Mustafa and Wrathall 2011). Table 2.1 shows the minimum lump sum allocations across the cropping seasons.

Any surplus waters in a given year are distributed according to the following percentages:

- Punjab 37 percent
- Sindh 37 percent
- Balochistan 12 percent
- NWFP (Khyber Pakhtunkhwa) 14 percent

The Indus River System Authority (IRSA) was set up by the Accord of 1991 to manage provincial water demands for reservoir releases and distribution to canal commands. The "Council of Common Interests," was introduced in the 1973 Constitution and reconstituted in 2009. It takes up disagreements among the provinces. IRSA does not have effective structures or mechanisms for regulating its political representation and technical administrative roles; the former is sometimes perceived to dominate the latter. In 2010, IRSA faced rising tensions leading to resignations and near-dissolution of its membership. Provinces have full authority to allocate their apportioned waters to various canal commands within their boundaries, which they do on a 10-daily operating basis. Few major canal commands cross provincial boundaries (though significant river flows, groundwater discharge, and drainage flows do cross provincial boundaries). While the Accord provides for excess flows and redistribution within provinces, it does not provide for extremely low flow conditions or negotiated transfers among provinces. A key analytical question is: How do

Table 2.1 Allocations per the Indus Accord

million acre-feet

Province	Kharif[a]	Rabi[b]	Total
Punjab	37.07	18.87	55.94
Sindh	33.94	14.82	48.76
NWFP[c]	3.48	2.30	5.78
Balochistan	2.85	1.02	3.87
Total	77.34	37.01	114.35

Source: Mustafa and Wrathall 2011.
Note: NWFP = North-West Frontier Province.
a. kharif period = April to September, spring planting season.
b. rabi period = October to March, winter planting season.
c. NWFP Civil canals = additional 3.00 million acre-feet.

these institutional constraints affect agricultural production patterns, values, and efficiencies?

Low Water-Use Efficiencies and Productivity

Low water-use efficiency and agricultural productivity are top concerns for the Government in Pakistan. Frequent comparison between low irrigated crop yields in Punjab, Pakistan, and Punjab, India (Ahmad 2005) find that both regions have lower yields compared to elsewhere (table 2.3). These comparisons are striking, but they are not as simple as they appear. Figure 2.10 shows that there has been slow growth in the overall trends in crop yields from 1991 to 2008 for all but maize. This may reflect a wide range of agronomic, economic, and technological factors. To what extent are yield differences based on water allocation differences, as compared with other inputs and resource conditions?

Low water-use efficiencies raise a comparable question. Efficiencies in the IBIS system comprise canal efficiencies, watercourse efficiencies, and field efficiency, measured as a percentage of water delivered relative to the amount withdrawn. When multiplied, they give a measure of system-wide water-use efficiency. Typical losses in Pakistan are shown in table 2.2.

Some irrigation scientists argue that subsequent reuse through pumping of canal seepage should be added, which would lead to higher estimates of system efficiency (Jensen 2007). Others argue for a shift from physical water-use efficiency to water productivity, measured either in terms of quantity of crop produced per cubic meter diverted and delivered, or in terms of the net caloric or economic value of that crop per unit of water (Molden et al. 2010). There are many ways to increase water productivity, from established techniques of watercourse improvement, precision leveling, and on-farm water management,

Figure 2.10 Trends in Crop Yields, 1991–2008

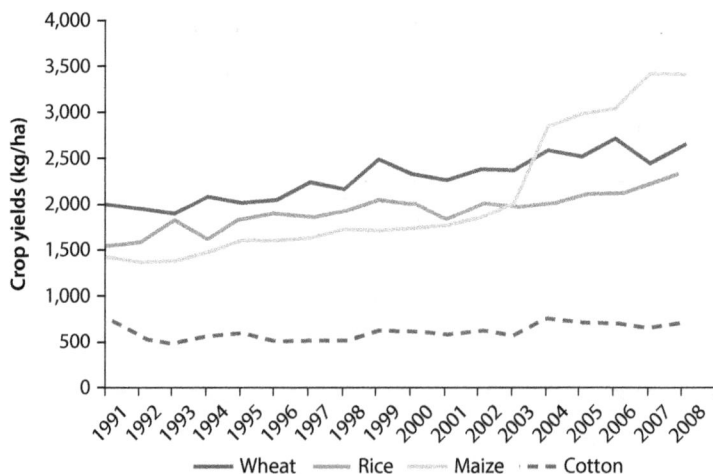

Source: Government of Pakistan, Ministry of Food, Agriculture, and Livestock 2010.

Table 2.2 Seepage Losses in Irrigation System

Location	Delivery at head, MAF	Losses percentage	MAF
Main and branch canals	106	15	16
Distributaries and minors	90	8	7
Watercourses	83	30	25
Fields	58	30	17
Crop use	41	n.a.	n.a.
Total		61	65

Source: GPCC 2005.
Note: n.a. = not applicable, MAF = million acre-feet.

to substitution of high-efficiency drip and sprinkler irrigation technologies for some crop and land types, as well as shifts to new crop types, varieties, and cultural practices. In a large system like the Indus, these alternatives have complex spatial as well as technological and economic linkages that need to be addressed through quantitative modeling.

National Policies and Plans on Water and Agriculture

National policies affect all sectors related to Indus Basin management. Of particular relevance are a recent constitutional change and a suite of long-term and short-term economic plans and budgets. These plans have had to address volatile economic and political conditions. During the past decade, Pakistan's real rate of gross domestic product (GDP) growth increased from 2 percent in 2001 to 9 percent in 2006 as a result of a combination of economic reforms, the end of a multiyear drought, and increased foreign funding related in part to the conflict in Afghanistan.

Increased growth reflected a combination of international and domestic factors. Pakistan took on a large international debt position during this period, which made it vulnerable to shocks such as the Kashmir earthquake in 2005, the global food price spike in 2008, and the ensuing economic recession, and Indus floods of 2010—all of which contributed to the drop in GDP growth rate to 2.7 percent by mid-2011 (World Bank 2012b). An IMF (2010) standby agreement extension strives to manage debt, in part through fiscal policies such as increasing tax revenues, privatization, and lowering subsidies. International economic pressures, coupled with domestic and international security problems, have eroded funding for water and agricultural development.

Constitutional Change

The 18th amendment to the Pakistan Constitution, passed in April 2010, eliminated the concurrent list of federal and provincial responsibilities and devolved most of the functions on that list to the provincial level. These functions include agriculture, including livestock and dairy; environment; and water management.

As an autonomous federal body, the Water and Power Development Authority (WAPDA) remains at the federal level, albeit with responsibilities limited to large water infrastructure planning, construction, and operations. As the Pakistan Meteorology Department (PMD) is under the Ministry of Defense, it also remains at the federal level. This constitutional change means that assessments of climate impacts and adaptation must devote increased emphasis on provincial planning, management, and governance. Further devolution of water management responsibilities to local government bodies has been attempted during the past decade and may resume in the future.

National Economic Long-Term Planning

The current long-term plan for Pakistan is titled *Vision 2030* (GPPC 2007). Its chapter on "Agricultural Growth: Food, Water and Land" includes major sections on agricultural production, water management, food security, and climate change—the first time this suite of sectors has been jointly addressed in a long-term planning document for Pakistan. *Vision 2030* begins with the observation that Pakistan has low rates of agricultural productivity, measured in yield-per-ha, compared with peer producers of food and fiber crops (table 2.3).

Vision 2030 proposes to address these gaps and minimize the impact of climate change in part by embracing the "gene revolution" (GPPC 2007, 53). Therefore, it is important to ask how crop breeding may affect water demand and, conversely, how hydroclimatic change could affect the productivity of new varieties. These uncertainties lie beyond current modeling capability and this report but may be an area for future investigation. At the same time, while *Vision 2030's* projected crop yields increase relatively steeply between 2005 and 2010, but flatten out over the next 20 years (table 2.4), the question is whether these targets are sufficient to meet food demands, given the future population demands and potential climate change impacts. This question is addressed in subsequent chapters on modeling.

Finally, the *Vision 2030* report—using the threshold of 1,000 m³/capita after 2010 and assuming a persistent high population growth rate—argues that

Table 2.3 Average Yields (kg/ha) of Selected Crops in Various Countries, 2005

Country	Wheat	Cotton	Rice (paddy)	Maize	Sugarcane
World	2,906	1,949	4,019	4,752	65,597
China	4,227	3,379	6,266	5,153	66,063
India	2,717	850	3,007	1,939	61,952
Egypt, Arab Rep.	6,006	2,603	9,538	8,095	121,000
Mexico	5,151	n.a.	n.a.	2,563	70,070
France	6,983	n.a.	n.a.	8,245	n.a.
Pakistan					
National Average	2,586	2,280	1,995	2,848	48,906
Progressive farmer	4,500	2,890	4,580	7,455	106,700

Source: GPPC 2007, 52.
Note: n.a. = not applicable.

Table 2.4 Crop Yield Targets of Major Agricultural Products
tons, millions

	Benchmark		Production targets	
Agricultural product	2004–05	2009–10[a]	2015[b]	2030[c]
Wheat	21.6	25.4	30.0	33.0
Rice	5.0	6.3	7.5	8.5
Cotton (lint)[d]	14.6	17.0	30.7	21.5
Sugarcane	45.3	56.7	63.4	n.a.
Fruits	6.0	7.0	10.8	n.a.
Oil seeds	5.8	7.5	8.12	n.a.
Meat	2.8	3.1	4.2	n.a.
Milk	29.4	43.3	52.2	n.a.
Fisheries	573.6	725	n.a.	n.a.

Source: GPPC 2007.
Note: n.a. = not applicable.
a. Mid-term development framework, 2005–10.
b. Ministry of Food, Agriculture, and Livestock, 2015.
c. Production based on regression analysis of 16 years of data (1990–2005).
d. bales, millions.

Table 2.5 Projected Sectoral Growth Rates during Mid-Term Development Framework Plan Period
percentage per year

Economic sector	Sectoral shares 2008–09	Sectoral growth rates						Average growth rate[a]	Sectoral share 2014–15
		2009–10	2010–11	2011–12	2012–13	2013–14	2014–15		
Agriculture	21.8	3.0	3.5	3.6	3.7	3.8	3.9	3.7	20.1
Industry	24.4	4.5	5.0	5.5	6.0	7.0	8.0	6.3	25.8
Services	53.8	3.2	3.6	4.8	5.2	6.2	7.4	5.4	54.1
GDP	100.0	3.5	3.9	4.7	5.1	5.9	6.8	5.3	100.0

Source: Panel of Economists 2010.
a. Growth rate during plan period 2010–11 to 2014–15.

an additional 12 MAF of storage is needed. This also incorporates the current observed reservoir sedimentation and future projections of increased general circulation model (GCM) monsoon rainfall of 20–30 percent.

Similarly, a panel of economists submitted recommendations for the next *Medium-Term Development Imperatives and Strategy for Pakistan* for a five-year period, 2010–2015.[5] The "Panel of Economists Final Report" (2010) envisions the agricultural growth rate to average only 3.7 percent, due in part to water constraints (table 2.5). The report urges increased irrigation efficiency, which it describes as averaging only 37 percent (due to canal, watercourse, and field losses). It criticizes the fiscal shortfalls of an irrigation revenue system (*abiana*) that recovers only 35 percent of its operation and maintenance costs. The report further recommends accelerated adoption of Bt cotton to emulate India's dramatic increase in yields in Bt cotton since 2002. The panel also advocates preparing for climate change, though it does not draw upon any current research or make specific policy recommendations.

Note that the previous mid-term development framework (MTDF) (2005–10) gave more detailed attention to physical water infrastructure investment. It cited the limited reservoir storage capacity in Pakistan, storage losses due to reservoir sedimentation, and irrigation seepage losses that are estimated to be 65 MAF or 61 percent of the water diverted into major canals (table 2.2). Moreover, it sought to lay out a comprehensive framework for water resources management, along with support for 36 continuing and 15 new water infrastructure projects, totaling more than PRs 276 billion over five years. The Agriculture chapter of the MTDF, by comparison, makes limited reference to issues of water management, which reflects a sector gap between irrigation and agricultural policy.

Current federal economic plans and budgets shed light on a number of policy issues relevant for addressing climate risks, water, and food security in the near term. There is increasing recognition of climate change issues in federal planning, but no climate policy has been included in an annual or five-year development plan or budget to date. A climate change strategy was approved by the Federal Cabinet in January 2012, and a new Ministry of Climate Change was created in March 2012 that could guide future planning and budgeting. The strategy gave heavy emphasis to adaptation in the water, agriculture and livestock, forestry, disaster preparedness, and vulnerable ecosystems (mountains, coastal zone, rangelands, wetlands), and human health. It also includes a mitigation section focused on energy, transport, and industries. Despite this emphasis, current development plans do not indicate where climate change risks would be addressed at the federal level.

The Pakistan Planning Commission needs to consider alternative agency and inter-agency organization for climate change policies and programs. The Planning Commission and Ministry of Finance will also need to consider the linkages between climate change and disaster risk reduction policies. Devolution of former federal sectoral functions to the provinces under the 18th Amendment will require stronger policy linkages between the federal water sector and provincial agricultural sectors. This will require vision and budget support at the federal level.

National Water Policies

The primary policy document in the water sector at the federal level is the WAPDA *Vision 2025*, developed in 2001, which describes WAPDA's long-term infrastructure development plan. Proposed water projects are described on the WAPDA web pages and are almost entirely physical infrastructure projects (no reference to climate change). The written report, updated in 2004, presents the overall context and rationale for these projects. WAPDA has prepared a "Developmental Plan" that focuses on strategic issues and infrastructure completed, planned, and phasing; it makes no reference to climate change. A major Water Sector Strategy was drafted in 2002, and adopted in 2005, but still remains in draft form. Thus, there is no strong policy linkage between WAPDA's

Vision 2025 for reservoir and hydropower infrastructure investment at the national level to increase storage and hydropower capacities, on the one hand, and the various provincial water sector policies that must address issues of water demand management and agricultural productivity, on the other.

National water policy is articulated in the Annual Plans of the Planning Commission and budgets of the Ministry of Finance. The most recent *Annual Plan 2011–2012* introduced the "Water Resources" sector as a balanced program of supply augmentation and irrigation management. The water sector plan and budget for 2010–11 had ambitious aims that had to be dramatically scaled back due to the 2010 flood and budget cuts. Quantitative targets in the two most recent plans indicated declining physical achievements and targets both before and after the 2010 flood. Although the plans indicated a partial shift from large projects to small- and medium-size projects, there are continuing efforts to advance Basha-Diamer and other large storage and hydropower projects central to WAPDA's *Vision 2025*.

Task Force on Food Security

The 2008 spike in world food prices led the Government of Pakistan to set up a Task Force on Food Security, which delivered its final report in 2009 (GPPC 2009). Its key points are that Pakistan needs to develop: (1) a national food security strategy (supported by 4 percent annual agricultural growth rate, efficient and equitable storage and pricing, increasing food access through a pro-poor growth and employment, and transparent safety nets); (2) a food security index for monitoring purposes; (3) favorable terms for agricultural trade and increased agricultural credit; (4) capacity-building in the federal [now provincial] agricultural departments; and (5) legislation in the form of a Seed Act Amendment Bill and Plant Breeders Rights Bill.

The task force recommendations focus on agricultural growth through increased yields, a shift to higher value horticultural crops, and increased investment in the high-value livestock and dairy sector. Attention was also given to enhancing agricultural water management and water-use efficiency through precision land leveling, watercourse improvements, water-efficient irrigation technologies, low delta water crops, and promotion of water saving technologies like drip and sprinkler irrigation.

The report identified water as a major constraint in agriculture. A serious structural and administrative barrier to achieving the production targets set in the report is the stressed irrigation system, which is stressed due to inadequate maintenance and inefficient water use that adversely affects the water balance. Water shortages are particularly severe in the southern part of the country where irrigation has expanded into some of the driest regions. The non-economic water prices also provide no incentive to adopt recommended cropping patterns and water-saving techniques. Although the national water policy provides a legal framework for water pricing and cost recovery to ensure effective and efficient water management, its implementation is poorly managed. The inefficient use of water was cited as one of the major issues in the comparably low levels of crop productivity—Pakistan's cereal production of 0.13 kg per cubic meter of irriga-tion water compares unfavorably with 0.39, 0.82, 1.56, and 8.7 kg in India, China, the United States, and Canada, respectively (Kumar 2003).

The task force further recommended a two-pronged strategy for the develop-ment of water resources to attain and sustain food security in Pakistan. First, attention should be paid to reducing water losses and improving conservation of available water resources to enhance productivity and increase cropping inten-sity. This task should include the continuity of ongoing development projects (watercourse improvement and high-efficiency irrigation systems), and expand-ing the coverage of new initiatives and pilot activities, such as laser land leveling and permanent raised bed, furrow irrigation. Second, new small-scale irrigation facilities in rainfed areas and poverty pockets of fragile eco-zones should be developed. It is estimated that the present cropped area of 58.5 million acres can be increased by at least 12 million acres from the available culturable wastelands of 20.6 million acres in the country.

Task Force on Climate Change

The Government of Pakistan Planning Commission set up a Task Force on Climate Change in October 2008 to provide appropriate guidelines for ensuring the security of vital resources such as food, water, and energy. Their final report, drafted in February 2010 (GPPC 2010), contributed to the formulation of a climate change policy that has been helping the Government pursue sustained economic growth by addressing the challenges posed by climate change. The report acknowledged the limited scope for expanding water supplies and advised that Pakistan would have to improve the efficiency of water use in all the sectors, particularly in agriculture. It also warned of the risk of increased demand of irrigation water because of higher evaporation rates at elevated temperatures in the wake of reducing per capita availability of water resources and increasing overall water demand. The report predicted that the impacts on food security in the agriculture sector would mainly be through reduced crop productivity caused by extreme events (floods, droughts, and cyclones). Given these risks under the increasing pressure of a growing population, Pakistan has no option but to take major steps for increasing its land productivity and water-use efficiency.

Notes

1. Includes 17 background papers.

2. Accessible online at http://web.worldbank.org/WBSITE/EXTERNAL/COUNTRIES/ SOUTHASIAEXT/0,contentMDK:20320047~pagePK:146736~piPK:583444~theSi tePK:223547,00.html.

3. http://siteresources.worldbank.org/SOUTHASIAEXT/Resources/223546-1171996340255/BagliharSummary.pdf.

4. For example, the Jang publishing group and Aman ki Asha sponsored a group of Indian and Pakistani leaders to discuss the prospects for international cooperation;

see http://www.amankiasha.com/events.asp in June 2010. Also the International Centre for Integrated Mountain Development convened a joint scientific meeting on the hydroclimatology of the upper Indus in 2010.

5. This medium-term planning timeframe is the functional equivalent of a five-year plan.

References

ADB (Asian Development Bank) and World Bank. 2010. "Pakistan Floods 2010 Damage and Needs Assessment." Pakistan Development Forum, Islamabad, November 14–15.

Ahmad, M., and G. P. Kutcher. 1992. "Irrigation Planning with Environmental Considerations: A Case Study of Pakistan's Indus Basin." World Bank Technical Paper 166, World Bank, Washington, DC.

Ahmad, S. 2005. "Water Balances and Evapotranspiration." Background paper for *Pakistan's Water Economy: Running Dry.* Washington, DC: World Bank and Oxford University Press.

Amir, P. 2005a. "Modernization of Agriculture." Background paper for *Pakistan's Water Economy: Running Dry.* Washington, DC: World Bank and Oxford University Press.

———. 2005b. "The Role of Large Dams in the Indus Basin." Background paper for *Pakistan's Water Economy: Running Dry.* Washington, DC: World Bank and Oxford University Press.

Bedi, Baba Pyare Lal. 2003. *Harvest from the Desert. The Life and Work of Sir Ganga Ram.* Lahore, Pakistan: NCA.

Berger, L. I., and IAC. 2011. "Preparation of Regional Plan for the Left Bank of Indus, Delta, and Coastal Zone for the Sindh Irrigation and Drainage Authority." http://www.sida.org.pk/download/lbg/phaseIII/Volume%20I%20&%20II%20-%20Draft.pdf. (accessed January 28, 2013).

Bhatia, R. 2005. "Water and Growth." Background paper for *India's Water Economy: Bracing for a Turbulent Future.* Washington, DC: World Bank and Oxford University Press.

Bhutta, M. N., and L. K. Smedema. 2005. "Drainage and Salinity Management." Background paper for *Pakistan's Water Economy: Running Dry.* Washington, DC: World Bank and Oxford University Press.

Blackmore, D. and F. Hasan. 2005. "Water Rights and Entitlements." Background paper for *Pakistan's Water Economy: Running Dry.* Washington, DC: World Bank and Oxford University Press.

Briscoe, J., and U. Qamar, eds. 2006. *Pakistan's Water Economy: Running Dry* (includes CD of background papers, 2005). Washington, DC: World Bank.

Falkenmark, M., A. Berntell, A. Jägerskog, J. Lundqvist, M. Matz, and H. Tropp. 2007. "On the Verge of a New Water Scarcity: A Call for Good Governance and Human Ingenuity." SIWI Policy Brief. Stockholm International Water Institute, Stockholm.

FAOSTAT (Food and Agricultural Organization of the United Nations Statistical Database). 2012. Database of Food and Agriculture Organization of the United Nations, Rome. http://faostat.fao.org/.

Gilmartin, D. 1994. "Scientific Empire and Imperial Science: Colonialism and Irrigation Technology in the Indus Basin." *Journal of Asian Studies* 53: 1127–49.

Government of Pakistan, Ministry of Food, Agriculture, and Livestock. 2010. *Agricultural Statistics of Pakistan 2008–2009.* Islamabad.

Government of Pakistan, Private Power and Infrastructure Board. 2011. *Hydropower Resources of Pakistan*. Islamabad. http://www.ppib.gov.pk/hydro.pdf.

GPPC (Government of Pakistan, Planning Commission). 2005. *Medium-Term Development Framework, 2005–2010*. "Chapter 27: Water Resources", Table 5. Islamabad. http://pc.gov.pk/mtdf/27-Water%20Sector/27-Water%20Sector.pdf.

———. 2007. *Vision 2030*, "Chapter 6: Agricultural Growth: Food, Water and Land." Islamabad. http://www.pc.gov.pk/vision2030/Pak21stcentury/vision%202030-Full.pdf.

———. 2009. *Final Report of the Task Force on Food Security*. Islamabad.

———. 2010. *Task Force on Climate Change Final Report*. Islamabad.

Hussain, I., R. Sakthivadivel, U. Amarasinghe, M. Mudasser, and D. Molden. 2003. *Land and Water Productivity of Wheat in the Western Indo-Gangetic Plains of India and Pakistan: A Comparative Analysis*. Research Report 65, International Water Management Institute, Colombo.

IMF (International Monetary Fund). 2010. *Pakistan: Fourth Review Under the Stand-By Arrangement, Requests for Waivers of Performance Criteria, Modification of Performance Criteria, and Rephasing of Access*. Staff Report; Staff Statement and Supplement; Press Release on the Executive Board Discussion; and Statement by the Executive Director for Pakistan. IMF Country Report 10/158. Washington, DC (accessed October 21, 2012). http://www.imf.org/external/pubs/ft/scr/2010/cr10158.pdf.

Jensen, M. E. 2007. "Beyond Irrigation Efficiency." *Irrigation Science* 25(3): 233–245.

Kumar, M. D. 2003. "Food Security and Sustainable Agriculture in India: The Water Management Challenge". Working Paper 60, International Water Management Institute, Colombo, Sri Lanka.

Michel, A. A. 1967. *The Indus Rivers: A Study of the Effects of Partition*. New Haven, CT: Yale University Press.

Molden, D., T. Oweis, P. Steduto, P. Bindraban, M. A. Hanjra, and J. Kijne. 2010. "Improving Agricultural Water Productivity: Between Optimism and Caution." *Agricultural Water Management* 97 (4): 528–35.

Mustafa, D., and D. Wrathall. 2011. "Indus Basin Floods of 2010: Souring of a Faustian Bargain." *Water Alternatives* 4 (1): 72–85.

Panel of Economists. 2010. *Final Report, Panel of Economists, Medium-Term Development Imperatives and Strategy for Pakistan*. Government of Pakistan, Planning Commission, Islamabad. http://pide.org.pk/pdf/Panel_of_Economists.pdf.

Siddiqi, A., J. Wescoat, S. Humair, and K. Afridi. 2012. "An Empirical Analysis of the Hydropower Portfolio in Pakistan." *Energy Policy* 50 (November): 228–41.

United Nations Population Division. 2012. "Population Prospects." http://esa.un.org/unpd/wpp/unpp/panel_population.htm.

Van Steenbergen, F., and M. S. Gohar. 2005. "Groundwater Development and Management in Pakistan." Background paper for *Pakistan's Water Economy: Running Dry*. Washington, DC: World Bank and Oxford University Press.

WAPDA (Water and Power Development Authority). 2004. *Water Resources and Hydropower Development Vision—2025 (Revised)*. Government of Pakistan Water and Power Development Authority, Lahore.

Wescoat, James L., Jr., S. Halvorson, and D. Mustafa. 2000. "Water Management in the Indus Basin of Pakistan: A Half-Century Perspective." *International Journal of Water Resources Development* 16 (3): 391–406.

World Bank. 2006. *Pakistan Strategic Country Environmental Assessment,* vol. 2, *Technical Annex: The Cost of Environmental Degradation in Pakistan—An Analysis of Physical and Monetary Losses in Environmental Health and Natural Resources.* Report 36946-PK. World Bank, Washington, DC.

———. 2012a. "Indus Waters Treaty." http://web.worldbank.org/WBSITE/EXTERNAL/COUNTRIES/SOUTHASIAEXT/0,contentMDK:20320047~pagePK:146736~piPK:583444~theSitePK:223547,00.html.

———. 2012b. *World Development Indicators Databank (WDI).* Online database. http://databank.worldbank.org/ddp/home.do?Step=12&id=4&CNO=2.

CHAPTER 3

Hydrology and Glaciers in the Upper Indus Basin

Key Messages

- Considerable speculation but little analysis exists concerning the importance of glaciers in the volume and timing of flow in the Indus River and its tributaries, as well as on the potential impact of climate change on these rivers.
- The two principal sources of runoff from the Upper Indus Basin (UIB) are (1) winter precipitation as snow that melts the following summer and (2) glacier melt. In the case of seasonal snow runoff volume, winter precipitation is most important. In the case of glacier melt volume, it is summer temperature.
- Using a simple model of these dynamics, it is estimated that glacier runoff contributes approximately 19.6 million acre-feet (MAF) to the total flow of the UIB, representing an estimated 18 percent of the total flow.
- The most probable source for a majority of the remaining 82 percent is melt water from the winter snowpack.
- Future runoff regimes will be determined primarily by changes in winter precipitation and summer temperatures.
- Given the orographic complexity of the region, general circulation model (GCM) projections are unlikely to have much value for forecasting purposes.
- There is a need for major investment in snow and ice hydrology monitoring stations, further scientific research, and forecasting to improve the hydrologic predictability of the UIB.

The mountain ranges encircling the Tibetan Plateau are a complex highland-lowland hydrologic system involving a range of water supply and use environments. The importance of the mountain contribution to the total flow of the major rivers of Asia, and the sources of runoff within individual mountain catchment basins, varies throughout the region. In addition to the limited studies of the general hydrology of the mountain catchments of these rivers, there are major issues of water use, as populations grow inexorably and many Asian countries begin a transition from agriculture-based systems to more industrialized economies.

Recent concerns related to climate change, retreating Himalayan glaciers, and the role played by these glaciers in the rivers of South Asia (for example, IPCC 2007; Rees and Collins 2004; World Wildlife Fund 2005) have served to illustrate how very little the scientific and water management communities understand about the role of the mountain headwaters (and glaciers in particular) to these river systems. The credibility of these concerns is in relation to several primary areas: (1) the contribution of glacier melt in the annual volume of stream flow; (2) the contribution of other sources, such as snowmelt and the summer monsoon; and (3) the credibility of climate change scenarios used to forecast future relationships in the complex terrain of the Hindu Kush–Himalaya mountain chain.

While there is a long history of scientific visits to the Karakoram Himalaya (Kick 1960), most have been primarily exploratory, resulting more in description than analysis. Much of the present understanding of the climate, hydrology, and glaciers of these mountains is based on a few analyses of a very limited data base. Archer et al. (2010) discussed the extremely limited number of climate stations in the Upper Indus Basin (UIB). In an area of over 160,000 km^2 above the Tarbela Reservoir, there are only 5 hydrometric stations in the main stem of the Indus River at the present time, and fewer than 20 manual climate stations. This compares with a total of 28 hydrometric stations and more than 250 climate stations in a comparable area in the Nepal Himalaya. Credible recent glacier mass balance data are available for few glaciers in the Karakoram, the Biafo, (for example, Hewitt 2010), and the Baltoro, (Mayer et al. 2006), and one, the Chhote Shigri Glacier, in the Chenab Basin in the western Himalaya (Wagnon et al. 2007). The most detailed analyses of climate data are a series of papers by Archer and his co-workers written during the period 2003–10. Glacier studies in these areas are largely the work of Hewitt and Young, and their students during several decades (Hewitt, 1968, 1998, 2005; Hewitt and Young 1993; Wake 1988, 1989), with more recent contributions by others (for example, Mayer et al. 2006; Wagnon et al. 2007).

There is no compelling evidence either for or against the impact of a changing climate on the hydrometeorology and glaciers of the UIB. Part of this is because there is a very limited database describing the climate and hydrology of these mountains, part has to do with the relative lack of familiarity of the climatological community with analyses of the three-dimensional mosaic of topo-climates within the extreme terrain of the UIB, and part from the fact that at least some of glaciers of the Karakoram are presently advancing (Bolch et al. 2012) rather than retreating, counter to the global trend. Additional scientific studies are clearly warranted as well as major investment in snow and ice hydrology-monitoring stations to improve the hydrologic understanding of the UIB.

The Indus River

The Indus River is an international river, with headwater tributaries in China (Tibet), India, Pakistan, and Afghanistan. The river originates north of the Great Himalaya on the Tibetan Plateau. The main stem of the river runs through the Ladakh district of Jammu and Kashmir and then enters the northern areas of Pakistan (Gilgit-Baltistan), flowing between the western Himalaya and Karakoram Mountains. Along this reach of the river, stream flow volume is increased by gauged tributaries entering the main river from catchments in the Karakoram Mountains—the Shyok, Shigar,[1] Hunza, Gilgit, and, in the western Himalaya, the Astore River (Hewitt and Young 1993), as well as ungauged basins on the north slope of the western Himalaya (Byrne 2009). Immediately north of Mt. Nanga Parbat, the westernmost of the high peaks of the Himalaya, the river turns in a southerly direction and flows along the entire length of Pakistan, to merge into the Arabian Sea near the port city of Karachi in Sindh province. Tributaries to this reach of the river from the western Himalaya are the Jhelum, Chenab, Ravi, and Sutlez Rivers, from the Indian states of Jammu Kashmir and Himachal Pradesh, and the Kabul, Swat, and Chitral Rivers from the Hindu Kush Mountains. The total length of the river is c. 3,180 km (1,976 miles [mi]). The river's total drainage area exceeds 1,165,000 km² (450,000 square miles [mi²]).

This chapter covers the mountain headwaters of the Indus River, commonly referred to as the UIB. The UIB is considered here to be the glacierized catchment basins of the western Himalaya, Karakoram, and northern Hindu Kush Mountains (map 3.1). The Hunza, Shigar, Shyok, the Gilgit Basin in the Karakoram Himalaya,

Map 3.1 The Mountain Catchment Basins of the Indus River

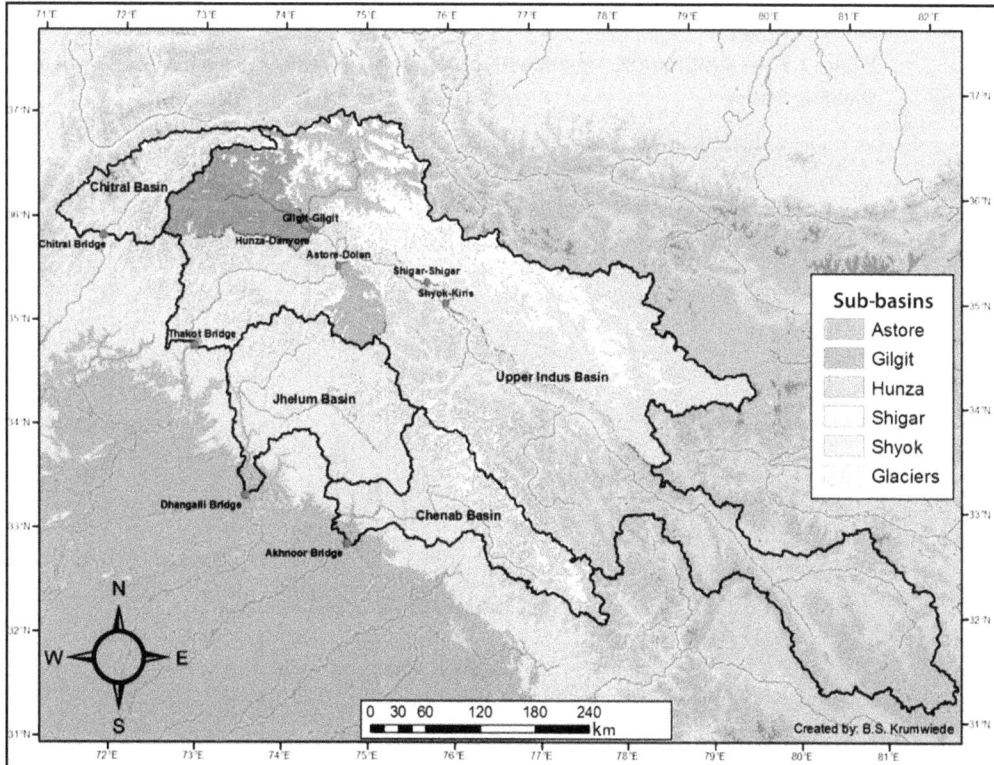

Note: The speckled blue area is the approximate area of glaciers and perennial snowfields. Gauging stations are represented by red dots.

and the Astore in the western Himalaya, contribute directly to the main stem of the Indus, with a total surface area of 166,065 km². The Jhelum and Chenab are tributaries from the western Himalaya, with a combined area of about 50,000 km², and the Chitral in the Hindu Kush Mountains extends approximately 12,000 km². Together these basins have a combined surface area of approximately 220,000 km² and contribute an approximately 110 MAF of the annual flow of the Indus River.

Within the mountain headwaters of the Indus River, the scale of vertical altitude differences and local relief has few analogues elsewhere in the world. Altitudes range from below 1,000 meters (m) where the river emerges on the plains at the two major controlling reservoirs of Tarbela and Mangla, to several mountain peaks above 8,000 m, including K2, the second-highest mountain on earth. As shown in figure 3.1, the mean altitude of the catchment above Besham, the gauging station immediately upstream from Tarbela Reservoir, is more than 4,000 m. This means that the greater part of the catchment surface is thrust up into the middle troposphere (ground level atmospheric pressures 700–500 millibars [mb]). The vertical lines in figure 3.1 represent atmospheric pressure levels often used by meteorologists as key heights for summary of circulation and weather processes. In lowland areas the behavior of climate variables, such as diurnal variations in air temperature, specific and relative humidity, wind

Figure 3.1 Area-Altitude Distribution (Hypsometry) of the UIB Catchment above Besham Gauging Station

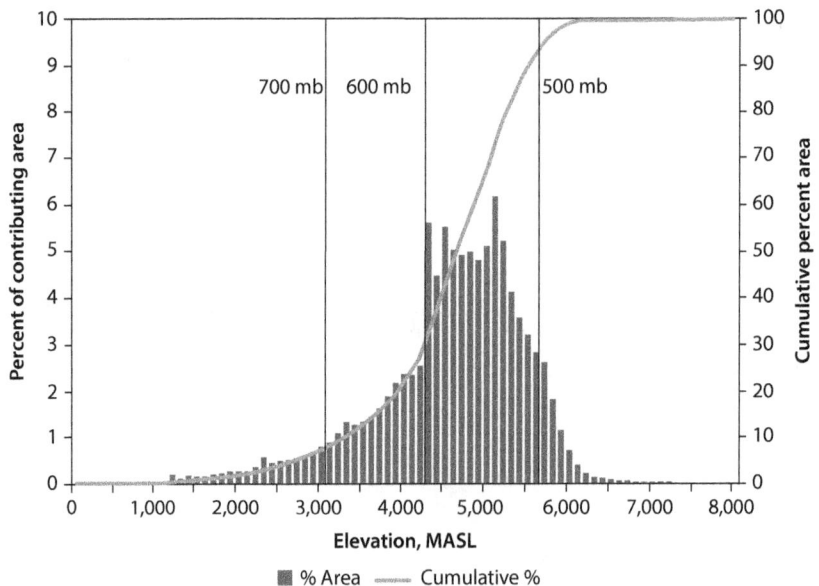

Source: © British Hydrological Society. Reproduced, with permission, from Forsythe et al. 2010; further permission required for reuse.
Note: MASL = meters above sea level, mb = millibars.

strength and direction, and cloud formation, are significantly different at these pressure levels than near to the ground surface.

Figure 3.1 is a graphical illustration of what may be a problem in the interpretation of most current climate change scenarios. While approximately 70 percent of the total surface area of the UIB above Besham is above the 600 mb level, the climate scenarios are generally more appropriate for altitudes considerably below the 700 mb level.

Hydrology of the Upper Indus Basin

Glaciers are a component of the hydrology of the mountain headwaters of this basin, and it is quite reasonable to expect that changes in the glaciers will be reflected in changes in the volume and timing of runoff from the mountain basins. The general hydrology of the Lower Indus Basin is assumed to be reasonably well-understood as learned from a network of gauging stations; reservoirs, such as the Tarbela and Mangla; and irrigation barrages on the piedmont. While this network provides data on which management decisions concerning water uses in the lower basin can be based, the hydrology of the upper basin remains largely a "black box." The general outlines of the hydrology of the UIB have been defined by several studies conducted in recent years, including Archer and Fowler 2004; Ferguson 1985; Goudie, Jones, and Brunsden 1984; Hewitt and Young 1993. The hydrology of the UIB has been described as having the following general characteristics:

- The mean annual flow of the UIB is approximately 58 MAF from the main stem above Tarbela Reservoir, 24 MAF from the Jhelum Basin, 22 MAF from the Chenab Basin, and 6 MAF from the Chitral Basin, for a total of 110 MAF.
- The total surface area of the main stem of the Indus above Tarbela is approximately 166,000 km^2, with an estimated glacier area of approximately 17,000 km^2. The other glacierized basin, the Chenab in the western Himalaya, has a surface area of 22,500 km^2 and a glacier area of 2,700 km^2.
- The two principal sources of runoff from the UIB are (1) winter precipitation as snow that melts the following summer and (2) glacier melt. In the case of seasonal snow runoff volume, winter precipitation is most important. In the case of glacier melt volume, it is summer temperature.
- Variability in the main stem of the Indus, based on the record from Besham, has ranged from approximately 85 to 140 percent of the period of record mean of 60 MAF.
- The wide diversity of hydrologic regimes in the mountain basins complicates the problem of relating stream flow timing and volumes to a uniform climate change.
- The mountain headwaters of the Indus River contribute approximately 60 percent of the mean annual total flow of the river, with approximately 80 percent of this volume entering the river system during the summer months of June–September.

The Indus Basin of Pakistan • http://dx.doi.org/10.1596/978-0-8213-9874-6

The Annual Hydrograph

Based on the mean period of record, stream flow begins to increase in May, with maximum runoff occurring in July in all sub-basins. This is consistent with what would be expected as the air temperatures increase and the freezing level migrates upward over the winter snow accumulation each spring. The July peak flow represents the end of snowmelt as a major source of surface runoff, as the winter snow deposit is removed by the rising freezing level. For Gilgit and Astore sub-basins, recession flow begins in July. This is interpreted as an indication that a glacierized area of 10 percent is not sufficient to produce a measureable stream flow volume. For the remaining gauged basins, all with glacierized surface areas greater than 20 percent, the summer runoff peak is maintained at a slightly lower volume through August, presumably by glacier melt. In early September, on average, the freezing level begins to migrate downward from near or slightly above 5,000 m. At this time each year, glacier melt ceases to be an important contributor to stream flow, and all runoff from the sub-basins enters the recession phase. Glacier melt becomes a component of stream flow, during a period of 1.0–1.5 months during August–September. The seasonality of both snowmelt and glacier melt for a specific basin appears to be determined by the area-altitude distribution of the basin, and varies among basins.

The Besham hydrograph, reflecting the combined contributions of all upstream sub-basins, shows a seasonal peak in July, assumed to represent peak snowmelt, but rather than beginning a recession phase at that point, has a secondary, slightly smaller, peak in August (figure 3.2). This is assumed to represent the glacier melt

Figure 3.2 Hydrograph Showing Mean Monthly Runoff per Year at Besham

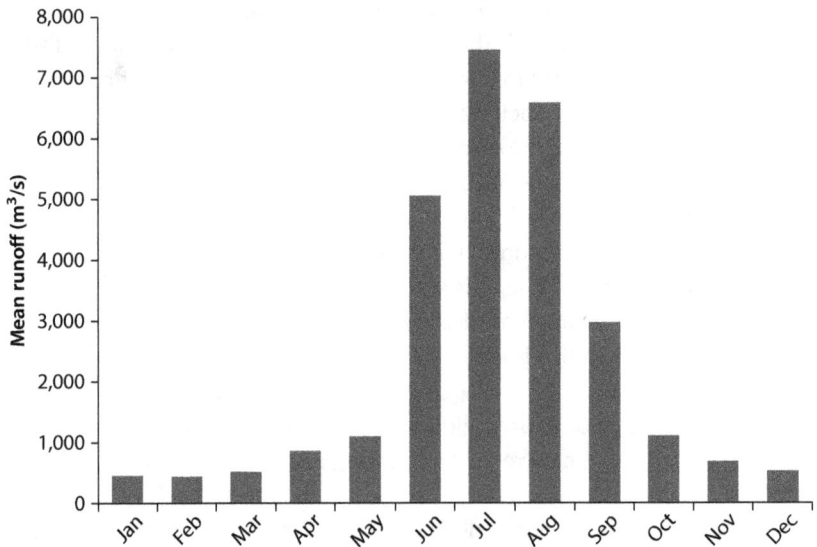

Source: WAPDA (unpublished data).
Note: Besham is a gauging station located immediately upstream from the Tarbela Reservoir on the main stem of the Indus River.

Figure 3.3 Annual Hydrographs of Gilgit and Hunza Basins

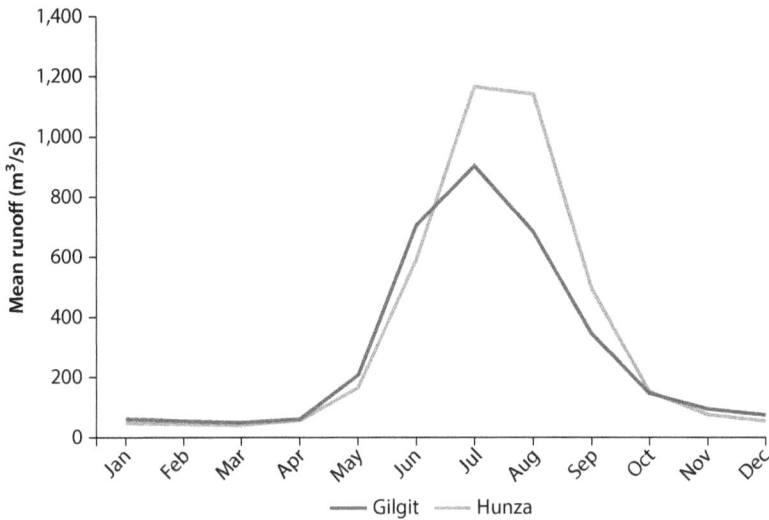

Source: WAPDA (unpublished data).

component of the annual stream flow. Following this second peak, the expected exponential recession curve begins.

For individual gauged basins in the UIB, the annual hydrograph is considered a good indicator of whether monthly runoff is primarily from melting winter snow deposit or glacier melt. This is illustrated by the annual hydrographs of the Gilgit and Hunza Basins (figure 3.3). The annual hydrographs of the Gilgit Basin (solid) and the Hunza Basin (dashed), illustrate the general difference in monthly flow volumes for a predominantly snow-fed basin and a basin with runoff resulting from both snowmelt and glacier melt. The two basins are almost equal in surface area, approximately 12,000 and 13, 000 km^2, respectively, and differ only slightly (about 8–10 km^3) in total annual discharge volume. Where they are most different is in glacier area. The Hunza has about 5,800 km^2 of glaciers, while the Gilgit has about 1,200 km^2. Both hydrographs are similar in shape, with a July maximum, the primary difference being that the Gilgit Basin has slightly higher volumes in the early spring and a peak flow in July, while the Hunza has much higher flow during both July and August and a higher volume in the early fall, suggesting a source of melt water beyond the winter snow.

Glacier Climates of the Upper Indus Basin

The literature provides several descriptions of the climates of the UIB. Thayyen and Gergan (2009) describe the geography of the hydrometeorological environments; Archer et al. (2010) describe the seasonality and altitudinal distribution of precipitation and temperature; and Hewitt (2010) provides a meteorological interpretation of the glacier climates. Glaciers can be found in all large mountain ranges, and they grow or shrink in response to the interaction between a regional climate and the topography of the mountains. The regional climate is modified by the topography of the mountains into a three-dimensional environmental mosaic, referred to as "topoclimates" (Thornthwaite 1953). The two most important topographic factors are altitude, and aspect. Altitude influences the physical properties of the air mass surrounding the mountains, primarily as a result of decreasing atmospheric density with increasing altitude. Aspect—the direction faced by mountain terrain—from a macro-slope of an entire mountain range to a cirque wall within that mountain range, influences the angle at which an air mass moving through the region intersects the mountain terrain, creating windward and leeward slopes. Aspect also is a major factor in determining the amount of solar radiation received at a surface. Solar radiation is the primary source of energy at higher altitudes in mountain ranges. There will be major differences in energy available for north- and south-facing slopes, largely unrelated to the mean air temperatures measured in adjacent valley floors.

Glaciers grow or shrink as a result of complex interactions between the processes of mass gain—in the form of snow—and energy exchange, primarily as short- and long-wave radiation and sensible heat. These interactions determine the mass balance of a glacier. The snow deposited annually, or seasonally, on the surface of a glacier represents a heat sink. When snow deposited on the glacier exceeds the amount of snow and ice that is removed by the annual amount of energy input, the mass balance is said to be positive, and over time the glacier will grow and advance. When the energy received is sufficient to melt both the annual snow deposits and the ice formed from snow deposits of previous years, the mass balance of the glacier is negative, and the glacier will retreat. Glaciers may advance or retreat from either an increase or decrease in energy availability, an increase or decrease in snow accumulation, or some combination of the two.

The average summer altitude of the 0°C isotherm, at which sufficient snowmelt and ice melt is possible to produce measureable runoff from a basin, is estimated to be approximately 5,000 m. A few valley glaciers in the Karakoram Himalaya have terminal altitudes below 3,000 m. At this altitude, ice melt is assumed to be occurring during most months of each year. This formation represents a very small fraction of the glacier cover of the UIB, however, and produces only an insignificant amount of runoff. The primary altitude of runoff volume produced by ice melt is immediately below the annual freezing level, where a combination of energy exchange and glacier surface area is maximized. In assessing the role of glacier melt in the rivers of South Asia, it is useful to remember that, presently, there are altitudes above approximately 5,000 m above which snow is deposited and never melts under present-day conditions. These glaciers exist through a range of altitudes from the lowest, where melt occurs continuously throughout the year, to the highest, where melt never occurs.

As inferred from the hydrological data, the hydrometeorology of the Karakoram tributaries to the main stem of the Indus River is dominated by a winter snowfall regime, with maximum snow-water equivalent (SWE) depths centered at approximately 4,000 meters above sea level (MASL). Between approximately 3,000 and 5,000 m, this snow melts each spring and summer and forms the bulk of the surface runoff. Following removal of the seasonal snowpack, glacier melt begins at these same altitudes and continues until all melt ceases in September. Above 5,000 m, there appears to be a rapid decrease in precipitation depth and glacier melt with altitude. Snowfall above 5,000 m is presumably redistributed by wind or avalanches into the topographic basins that form the accumulation zones of the glaciers. As a result of plastic flow, this snow is ultimately transferred to the altitude of the ablation zone of the glaciers at 3,000–5,000 MASL where it becomes the source of much of the August–September stream flow. In the western Himalaya basins of the Jhelum and Chenab Rivers, the winter snow is augmented by the summer monsoon, and, in the Chenab, by a small glacier melt component.

Distributed Process Models of Glaciers and Total Basin Runoff

The approach described here uses a very simple physical distributed process model, which is based on the assumption that, as a useful first approximation, the most important controls on the water budget of a mountain basin in the Hindu Kush-Himalayan Mountains are the altitudinal range occupied by the basin and the distribution of surface area within the basin. Altitude is used as a proxy for all major topographic variables—altitude, aspect, and slope—and temperature for both sensible heat and radiation, as exemplified by the use of the "degree-day" index (Ohmura 2001). Surface area is necessary to convert the specific values to total volumes. The areal distribution of runoff may be derived as the product of the area-altitude hypsometry of an entire catchment basin, or of selected portions such as the glacierized area of the basin, and the altitudinal gradient of the water budget over that portion of the basin. Much of the procedure is based on the application of traditional budget analysis procedures from hydrology or glaciology. Ideally, the basin should have a gauging station at its outlet, to provide an empirical test of the volume and timing estimates.

The Catchment Basins

A digital elevation model (DEM) was produced of the entire region occupied by the UIB from Shuttle Radar Topography Mission (SRTM) 90 m data. The perimeter of the entire basin to be included was determined, together with each of the individual gauged sub-basins within this basin. Catchment basins were defined as the drainage area upstream from a hydrometric gauging station. Basin boundaries above the stations were defined using the Watershed tool in the Hydrology toolset of Spatial Analyst Tools in ArcGIS 9.3.1 to define basin boundaries. The rasters were converted to polygon shape files, combining the basins and sub-basins, and the basin surface areas calculated (in km²). The results for all the basins included in this study are shown in table 3.1 and figure 3.4.

Table 3.1 illustrates the concentration of surface area at altitudes 4,000–6,000 MASL for many basins. The primary importance of this concentration of

Table 3.1 UIB Catchment Basins with Total Areas and Area-Altitude Distribution

1,000 m increments, km²

Station	0–1 k	1–2 k	2–3 k	3–4 k	4–5 k	5–6 k	6–7 k	7–8 k	8–9 k	Total
Thakot	240	3,305	9,443	26,110	68,278	56,493	2,726	111	1	166,707
Besham	172	3,083	9,212	26,028	68,274	56,490	2,725	111	1	166,096
Partab	0	644	4,809	19,150	62,015	56,224	2,677	99	1	145,618
Kachura	0	0	1,947	11,752	48,337	51,046	2,153	52	1	115,289
Kiris	0	0	477	2,785	8,337	20,141	1,588	22	0	33,350
Shigar	0	0	417	1,094	2,968	2,157	254	31	1	6,922
Danyore	0	138	848	2,632	5,620	3,997	454	44	0	13,732
Gilgit	0	179	1,246	3,534	6,832	875	15	0	0	12,680
Doian	0	23	336	1,489	1,985	134	18	3	0	3,988
Dhangalli	1,182	8,085	7,632	7,217	2,986	20	0	0	0	27,122
Aknoor	874	2,718	4,078	4,935	6,719	3,162	19	0	0	22,504
Chitral	0	156	1,505	3,490	5,398	1,769	173	14	0	12,505
Total	2,468	18,331	41,950	110,216	287,749	252,508	12,802	487	5	726,513

Figure 3.4 Upper Indus Basin Hypsometries of Table 3.1

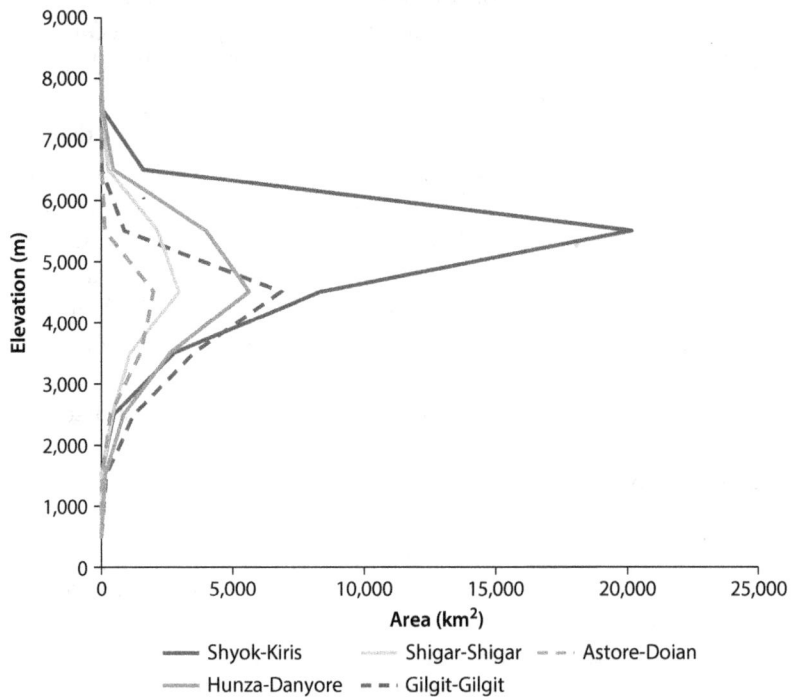

surface area at these altitudes is that it provides an extensive platform for the deposition of the winter snowfall. Beginning in the early spring, the freezing level gradually rises to the upper portion of this altitudinal belt, providing a large fraction of the summer-season stream flow volume. The area-altitude distribution of the hydrologic characteristics of the UIB is fundamental to a realistic assessment of the potential effects of climate change on the volume and timing of stream flow from the basin. While most gauged basins have a concentration of surface at 5,000 MASL, the Shyok Basin has a maximum concentration at 6,000 MASL. This suggests that the Shyok Basin, including a portion of the Baltoro Mustagh, may have an ice balance that is slightly more positive.

The Orographic Runoff Gradient

The gradient of total basin water budget with altitude was estimated from the relationship between the measured mean specific annual runoff (mm) and the mean altitude of the gauged basin (m). A curvilinear relationship between specific runoff and mean basin altitude is observed, with a maximum at 3,000–4,000 m and a minimum at the highest and lowest altitudes. It is assumed this distribution is produced by monsoon rain, as the encroaching summer monsoon is forced to rise over the Himalayan wall. Variation in the curvature of the gradient is assumed to be a result of a weakening of the summer monsoon as it moves from east to west along the Himalayan front. Estimating the orographic runoff gradient for the Karakoram Himalaya, in the UIB is more difficult. There are far fewer gauged basins in the Karakoram than in the Nepal Himalaya, and the range of mean altitudes of those basins is much narrower. To define the general form of the orographic gradient for the western Himalaya and Karakoram, specific runoff values and mean altitudes shown in table 3.2 were combined with similar data from winter snowpack SWE (from Forsythe et al. 2010) and the Karnali Basin, from western Nepal. The result is shown in figure 3.5. The data from snowpack SWE data from Forsythe et al. (2010). are shown in white, the Karnali Basin in eastern Nepal in gray, and the Karakoram basins and the western Himalaya tributaries to the UIB are in black. This data suggests that above 5,000 m there is negligible runoff being produced.

Glacier Melt and the Ablation Gradient

Haefeli (1962) postulated the existence of an "ablation gradient" to summarize the trend of melt from all processes with altitude over the ablation zone of a glacier (figure 3.6). In plotting data from reports in the literature, the author

Table 3.2 Basic Descriptive Statistics of the Basins in This Study

River	Sub-basin	Gauge site	Specific runoff (m)	Average altitude (m)
Indus	Astore	Doyan	1.29	3,981
	Gilgit	Gilgit	0.62	4,056
	Hunza	Danyore	0.76	4,516
	Shigar	Shigar	0.98	4,611
	Shyok	Kiris	0.32	5,083
	Indus	Besham	0.44	4,536
	Chitral	Chitral	0.71	4,120
Jhelum		Dhangalli	1.08	2,628
Chenab		Aknoor	1.22	3,542

Figure 3.5 Orographic Runoff Gradient for the Western Himalaya and Karakoram Sub-Basins

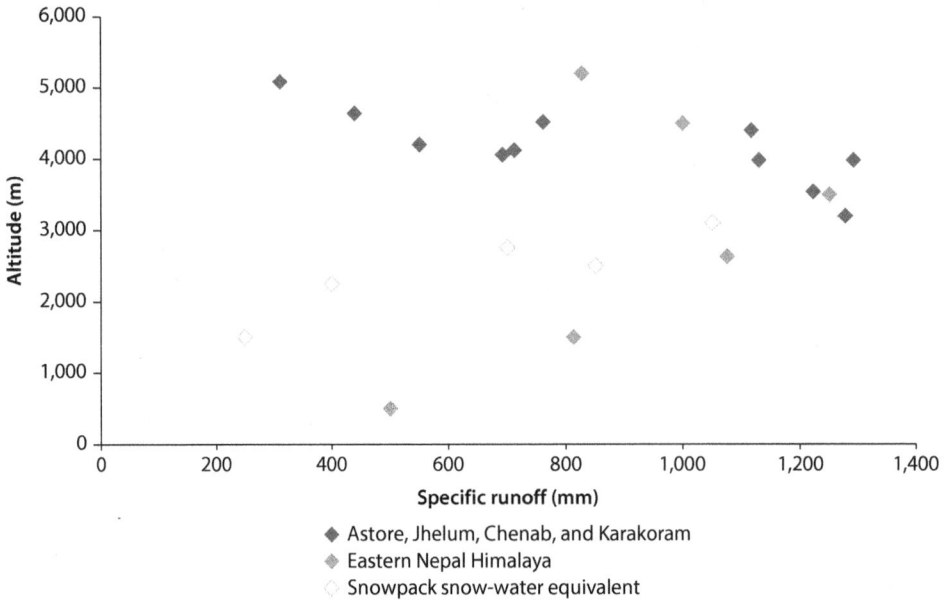

♦ Astore, Jhelum, Chenab, and Karakoram
♦ Eastern Nepal Himalaya
◇ Snowpack snow-water equivalent

Source: Based on data from sub-basins in the Astore, Jhelum, Chenab, and Karakoram, the eastern Nepal Himalaya, and snowpack SWE values from Forsythe et al. 2010.

Figure 3.6 The Ablation Gradient

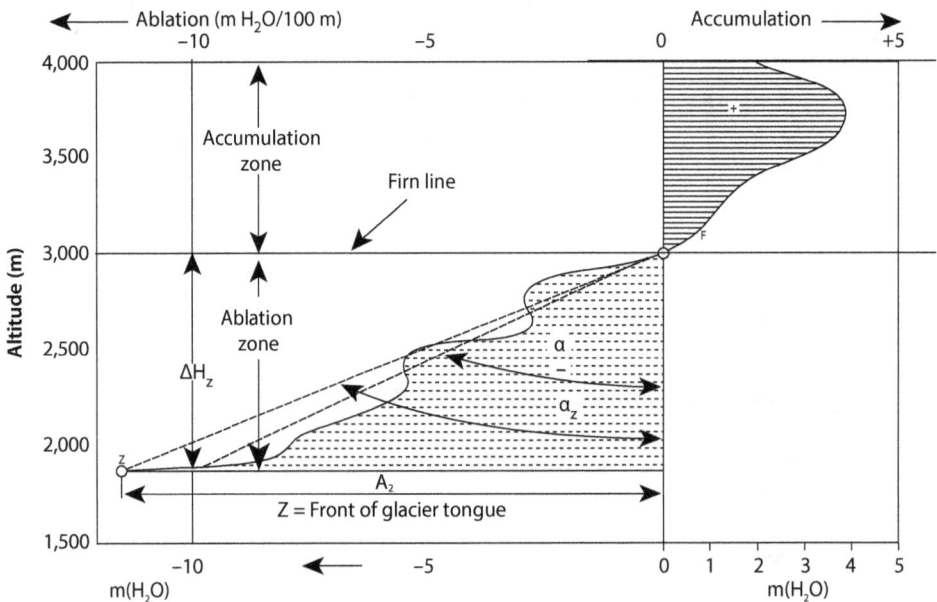

Source: © International Association of Hydrological Sciences (IAHS). Reproduced, with permission, from Haefeli 1962; further permission required for reuse.

found an inverse correlation in the slope of the ablation gradient with latitude, progressing from values of 0.2 m/100 m for glaciers in the high arctic to approximately 1 m/100 m at the latitude of the Karakoram Himalaya. According to Haefeli, "The ablation gradient is analogous to the well-known gradient of the average annual temperature of the air. The analogous phenomenon in the ablation would mean that the ablation gradient for a given glacier within a given climatic period remains approximately independent of the yearly fluctuations of the firn line" (50).

For the present study, an ablation gradient of 1m/100 m was assumed, based on studies of glaciers in the western Himalaya and Karakoram by Mayer et al. (2006) and Wagnon et al. (2007) (figure 3.7). Hewitt et al. (1989) estimated an ablation gradient of 0.5 m/100 m for the middle portion of the ablation zone on the Biafo glacier but did not present actual measurements.

The use of the ablation gradient concept requires that an altitude above which no ablation and runoff occurs be defined. For this study, this altitude is defined as the mean summer-season altitude of the 0°C isotherm. The mean altitude of the 0°C isotherm will be located at some intermediate altitude between that of

Figure 3.7 Four Years of Mass Budget Variation with Altitude, Chhota Shigri Glacier, Chenab Basin, Western Himalaya

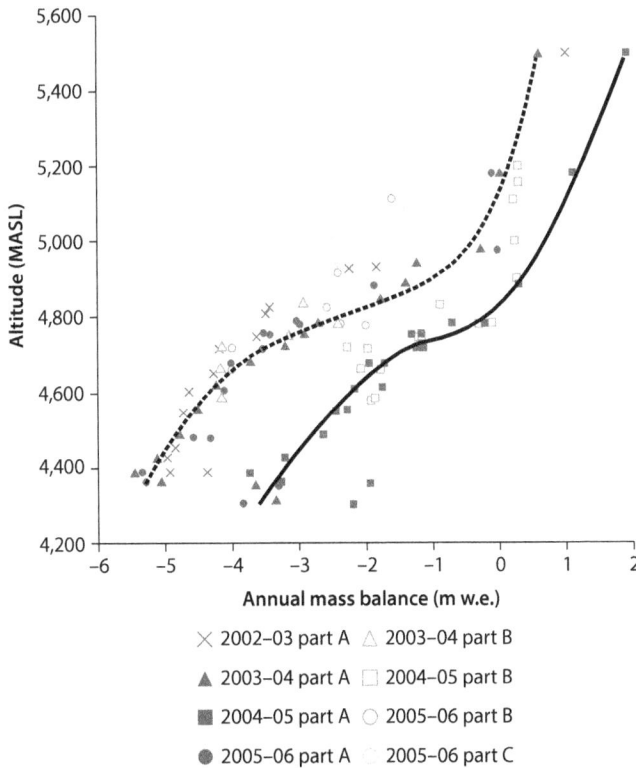

Legend:
- × 2002–03 part A
- △ 2003–04 part B
- ▲ 2003–04 part A
- □ 2004–05 part B
- ◼ 2004–05 part A
- ○ 2005–06 part B
- ● 2005–06 part A
- 2005–06 part C

x-axis: Annual mass balance (m w.e.)
y-axis: Altitude (MASL)

Note: MASL = meters above sea level, m w.e. = meters water equivalent.

Figure 3.8 Elevation of the Freezing Level for Monthly Maximum and Minimum Temperatures, Karakoram Himalaya

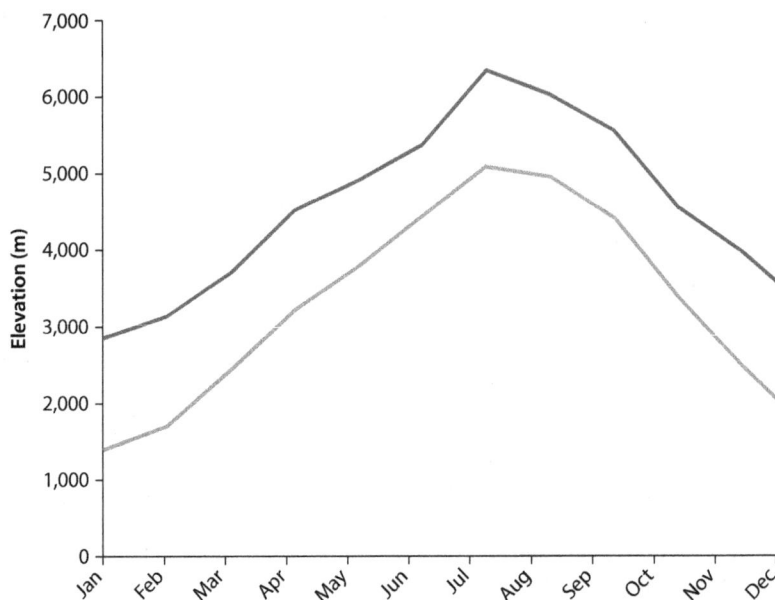

Source: © Archer and Fowler. Reproduced, with permission, from Archer and Fowler 2004; further permission required for reuse.

the minimum and maximum temperatures, as shown in figure 3.8. The estimates of glacier melt volume in this report are based on a summer-season freezing level of 5,000 m, above which some melt may occur but there is no measureable runoff. This level may be somewhat higher, on average, or may vary with location within the UIB. Any change in the altitude of the freezing level will have a considerable impact on the calculated volume of glacier melt and runoff, since the altitude of the freezing level is also the altitude of the maximum surface area belt of the glaciers.

The Estimated Glacier Component of Stream Flow

Values for each 100 m belt were determined from the ablation gradient, and the total ice melt was calculated as the sum of the product of the surface area of the respective belt and estimated ablation at that altitudinal interval. These values, summed for all the altitudinal belts on the ablating portion of the glaciers, were assumed to represent the annual ablation balance for the combined glaciers of each catchment basin. An assumed summer-season freezing level of 5,000 m and an ablation gradient of 1 m/100 m are used. The estimate of glacier melt to total stream flow in the UIB is based on a corrected surface area derived from an initial measurement of glacier surface area prepared by the National Snow and Ice Data Center (NSIDC) at the University of Colorado. This approach allows the calculation of the relative contribution of glacier melt and snowmelt as components in the annual flow of the UIB (table 3.3, figure 3.9). Results show that glacier

Table 3.3 Estimated Contribution of Glacier Melt and Snowmelt to Total Runoff for UIB Sub-Basins

Basin	Area, (km²)	Glacier, (km²)	q (mm)	Q (MAF)	Ice melt (MAF)	Snowmelt (MAF)
Hunza	13,734	4,339	0.76	8.5	4.0	4.5
Astore	3,988	450	1.29	4.2	0.8	3.4
Shigar	6,922	2,885	0.98	5.5	2.9	2.7
Shyok	33,350	6,221	0.32	8.7	4.9	3.8
Gilgit	12,682	994	0.62	6.4	1.5	4.8
Kachura (estimated)	75,000	n.a.	0.21	12.9	n.a.	12.9
Ungauged (estimated)	20,000	n.a.	0.72	11.8	n.a.	n.a.
Besham[a]	166,096	14,889	0.44	58.0	14.1	32.0
Chitral	11,396	2,718	0.71	6.6	3.2	3.4
Chenab	22,503	2,708	1.22	22.2	2.3	19.9
Jhelum	27,122	0	1.08	23.6	0	23.6
Total[b]	199,995	20,315		110.4	19.6	79.0

Note: n.a. = not applicable, MAF = million acre feet.
a. Ice melt and snowmelt contributions do not sum to the total flow (Q) because of unknown contributions from a 20,000 km² area. No glaciers are observed in this area, so it is likely that the remainder flow will be from either snow or the monsoon.
b. Total represents the sum of the Besham, Chitral, Chenab, and Jhelum basins.

Figure 3.9 Estimated Stream Flow Sources for the UIB Primary Glacierized Sub-Basins

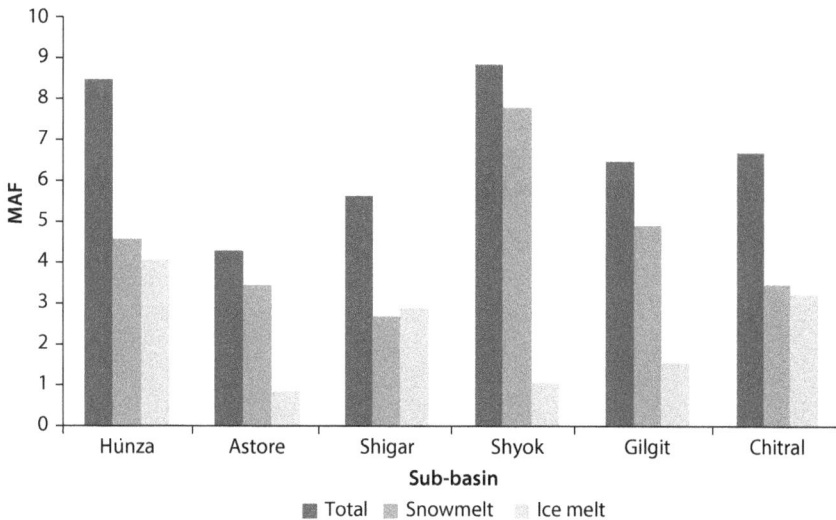

Note: MAF = million acre-feet.

runoff contributes approximately 19.6 MAF to the total flow of the UIB: 14.1 MAF from the Karakoram Himalaya, 2.3 MAF from the western Himalaya, and 3.2 MAF from the Hindu Kush. This represents an estimated 18 percent of the total flow of 110 MAF from the mountain headwaters of the Indus River. The most probable source for a majority of the remaining 82 percent is melt water from the winter snowpack.

Climate and Stream Flow Variability in the Upper Indus Basin

A reasonable concern is how much will a changing climate cause changes in the volume or timing of stream flow in the Indus River. Most scenarios of the impact of climate change on the hydrology of glacierized mountains have been based on the assumption that increasing air temperatures will produce an initial period of flooding, followed by an increasing drought as the glaciers retreat (Rees and Collins 2004). At least implicitly, such scenarios assume that current annual discharge volumes are relatively constant from year to year and that stream flow volume is primarily a result of glacier melt. The findings of this analysis based on analyses of the hydrographs from both glacierized and non-glacierized basins in the UIB do not provide support for either of the assumptions. This chapter demonstrates that snowmelt is the main source of annual stream flow to the UIB. Moreover, interannual variability may be determined, in part, by year-to-year fluctuations in both winter precipitation, as snow, and summer-season snowmelt and ice melt, as a result of fluctuations in energy availability. Some insight may be provided by an analysis of the variability of stream flow in the river under existing climate conditions.

The annual variation in stream flow in the main stem of the UIB (where roughly 80 percent of the glaciers of the entire basin are located) ranges from 140 to 80 percent of the mean. The variation is not symmetrical with respect to the long-term average volume (figure 3.10).

Approximately 70 percent of the annual flow from the sub-basins of the UIB occurs during July and August each year. These are months of maximum snowmelt (July) and glacier melt (August), as discussed earlier. An inspection

Figure 3.10 Percent Variation from Mean Annual Stream Flow at Besham, 1969–97

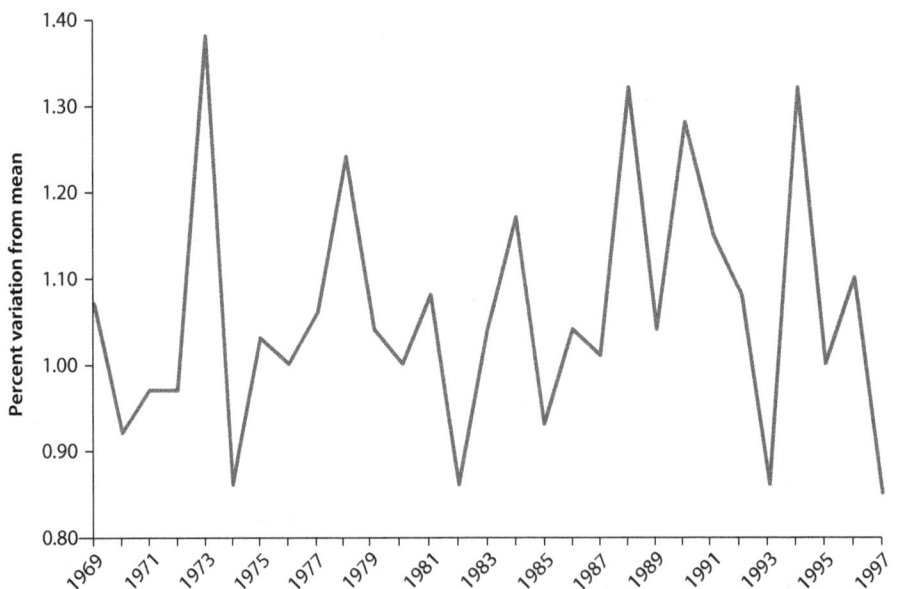

of the period-of-record summer-season runoff shows that the peak flow month varies from year to year, the frequency of this shift varying among basins, presumably as a result of variations from wet-cold to dry-warm conditions, increasing or decreasing the relative contribution of either snow-melt or glacier melt.

The peak annual flow times for several UIB sub-basins are as follows:

- For Besham, a basin with approximately a 15 percent glacier-covered area, the annual peak flow has occurred 75 percent of the time in July, and 25 percent of the time in August during the period of record (figure 3.11).
- For the Hunza Basin, with a glacier covered area of approximately 50 percent, the peak annual flow has occurred 60 percent of the time during August (figure 3.12).
- The annual peak flow from the Astore Basin, with approximately 10 percent glacier covered area, is consistently in July (figure 3.13).

These basins exemplify conditions in all gauged basins in the main stem of the UIB, illustrating the differences between the maximum and minimum glacierized areas in these basins. With a warming climate, it is assumed that there would be a shift to an increasing number of peak flows occurring in August; with a shift to a cooler-wetter climate, the July peak would become dominant.

For assessing the potential impact of climate change scenarios on stream flow in the UIB, it is useful to distinguish between those changes that could result from variations in precipitation from those related to changes in temperature. The volume of runoff from winter snow-melt will be determined primarily by variations in winter precipitation, since in all cases sufficient energy should be available during normal melt seasons to remove any realistic increases. On the other hand, glacier melt-water production will vary with the energy availability (change in temperature) during the melt season at the glacier surface. This also might not necessarily result from an increase or decrease in air temperature,

Figure 3.11 Summer Season and Annual Stream Flow in Besham Basin, 1970–95

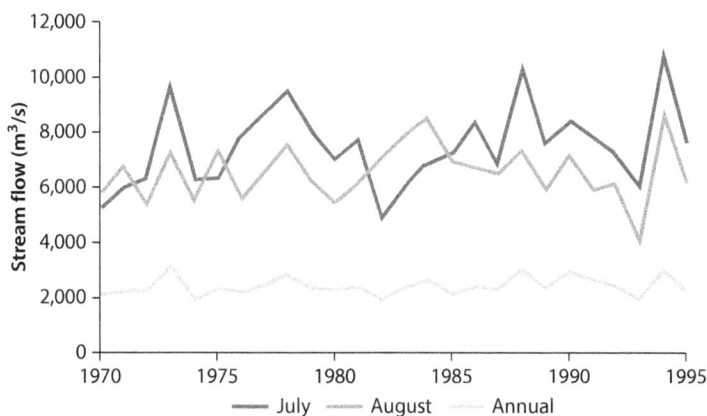

Figure 3.12 Summer Season Stream Flow in Hunza Basin (Significant Glacier Cover), 1966–96

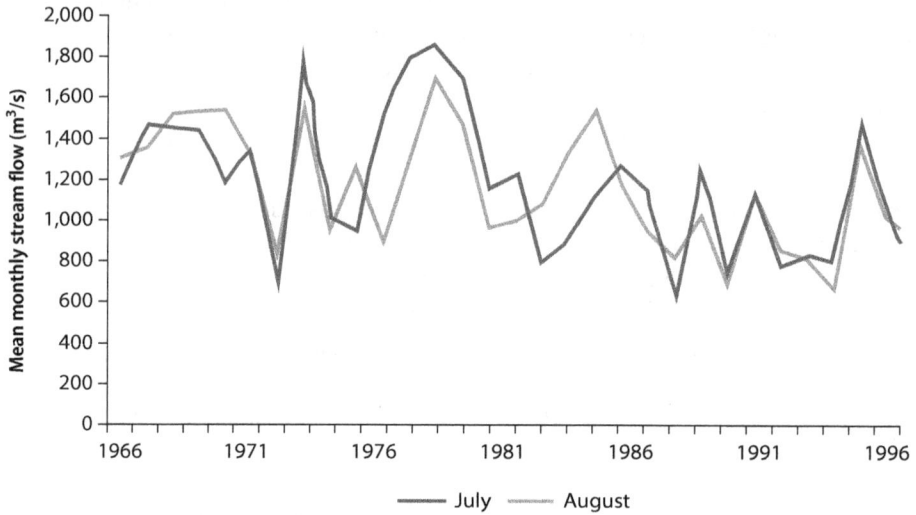

Figure 3.13 Summer Season Stream Flow in Astore Basin (Limited Glacier Cover), 1974–99

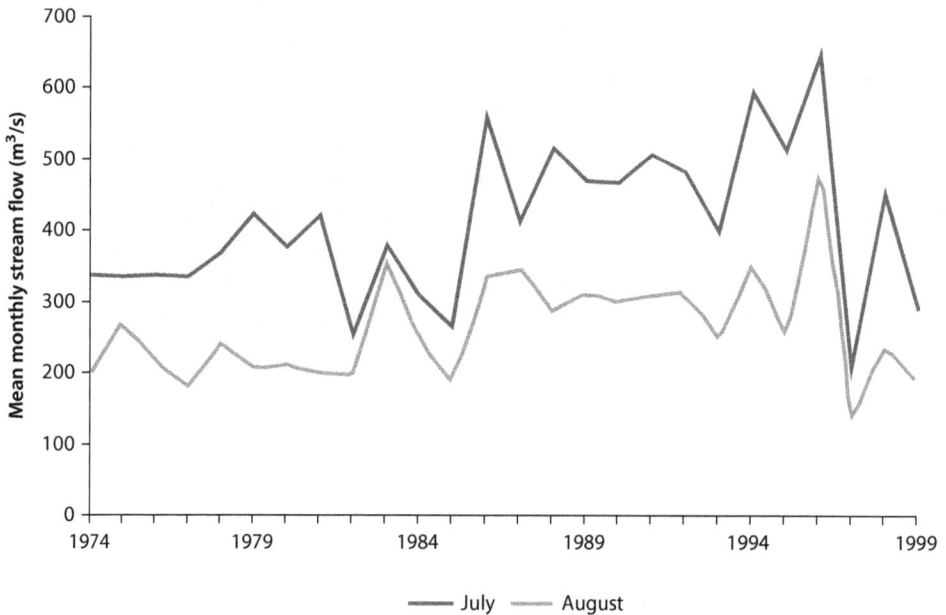

but could result from changes in summer cloudiness that increase or decrease receipt of shortwave radiation, or from the frequency of minor summer snow storms at the altitude of the glaciers that alter the albedo of the glacier surface.

Thus, the major challenge in predicting the impact of climate change on overall water resource availability in the UIB is to be able to make accurate predictions of changes (magnitude and direction) in winter precipitation and summer temperatures. This analysis also demonstrates that, since the large majority of total flow originates from snow, predictions of future precipitation change would be the top priority. Additional scientific studies, as well as major investment in snow and ice hydrology monitoring stations, will help to improve the hydrologic understanding of the UIB and future projections.

Note

1. The gauging station for the Shigar Basin has reportedly been discontinued (personal communication, D. Archer et al. 2010).

References

Archer, D. R., and H. J. Fowler. 2004. "Spatial and Temporal Variations in Precipitation in the Upper Indus Basin: Global Teleconnections and Hydrological Implications." *Hydrology and Earth System Sciences* 8 (1): 47–61.

Archer, D. R., N. Forsythe, H. J. Fowler, and S. M. Shah. 2010. "Sustainability of Water Resources Management in the Indus Basin under Changing Climatic and Socio Economic Conditions." *Hydrology and Earth System Sciences* 14: 1669–80.

Bolch T., A. Kulkarni, A. Kääb, C. Huggel, F. Paul, J. G. Cogley, H. Frey, J. S. Kargel, K. Fujita, M. Scheel, S. Bajracharya, M. Stoffel. 2012. "The State and Fate of Himalayan Glaciers" *Science* 336 (6079): 310–14.

Byrne, M. 2009. "Glacier Monitoring in Ladakh and Zanskar, Northwestern India." Unpublished master's thesis, Department of Geography, University of Montana, Missoula, MT.

Ferguson, R. I. 1985. "Runoff from Glacierized Mountains: A Model for Annual Variation and Its Forecasting." *Water Resources Research* 21 (5): 702–08.

Forsythe, N., C. Kilsby, H. Fowler, and D. Archer. 2010. "Assessing Climate Pressures on Glacier-Melt and Snow Melt-Derived Runoff in the Hindu Kush-Karakoram Sector of the Upper Indus Basin." Paper Presented at the British Hydrological Society Third International Symposium, "Managing Consequences of a Changing Global Environment," Newcastle University, U.K., July 19.

Goudie, A. S., D. K. C. Jones, and D. Brunsden.1984. "Recent Fluctuations in Some Glaciers of the Western Karakoram Mountains, Hunza, Pakistan." In *The International Karakoram Project*, Vol. 2, edited by K. J. Miller. London: Royal Geographical Society.

Haefeli, R. 1962. "The Ablation Gradient and the Retreat of a Glacier Tongue." *IAHS Publication* 58: 49–59.

Hewitt, K. 1968. "Records of Natural Damming and Related Events." *Indus* 10 (4): 4–14.

———. 1998. "Catastrophic Landslides and Their Effects on the Upper Indus Streams, Karakoram Himalaya, Northern Pakistan." *Geomorphology* 26: 47–80.

———. 2005. "Interactions of Rock Avalanches and Glaciers in High Mountains, with Particular Reference to the Karakoram Himalaya, Inner Asia." Abstr. EGU05-A-01195 (NH3.02-1FR20-004), 449. European Geosciences Union, General Assembly, Vienna.

———. 2010. "Glaciers and Climate Change in the Karakoram Himalaya: Developments Affecting Water Resources and Environmental Hazards." Unpublished manuscript. Cold Regions Research Centre, Wilfrid Laurier University, Waterloo, Ontario, Canada.

Hewitt, K., and G. J. Young. 1993. "Glaciohydrological Features of the Karakoram Himalaya: Measurement Possibilities and Constraints." In *Snow and Glacier Hydrology*. Proceedings of the Kathmandu Symposium, Nov. 1992. Association of Hydrological Sciences, IAHS/AISH Publication 218: 273–83. Oxfordshire, U.K.

Hewitt, K., C. Wake, G. J. Young, and C. David. 1989. "Hydrological Investigations at Biafo Glacier, Karakoram Himalaya, Pakistan: An Important Source of Water for the Indus River." *Annals of Glaciology* 13: 103–08.

IPCC (Intergovernmental Panel on Climate Change). 2007. *Synthesis Report: Contribution of Working Groups I, II and III to the Fourth Assessment Report of the Intergovernmental Panel on Climate Change*, edited by R. K. Pachauri and A. Reisinger. Geneva, Switzerland.

Kick, W. 1960. "The First Glaciologists on Central Asia, According to New Studies in the Department of Manuscripts at the Bavarian State Library." *Journal of Glaciology* 3 (28): 687–92.

Mayer, C., A. Lambrecht, M. Belo, C. Smiraglia, and G. Diolauti. 2006. "Glaciological Characteristics of the Ablation Zone of Baltoro Glacier, Karakoram, Pakistan." *Annals of Glaciology* 43 (1): 123–31.

Ohmura, A. 2001. "Physical Basis for the Temperature-Based Melt-Index Method." *Journal of Applied Meteorology* 40 (4): 753–61.

Rees, G., and D. Collins. 2004. *SAGARMATHA: Snow and Glacier Aspects of Water Resources Management in the Himalayas*. DFID Project R7980—An Assessment of the Potential Impacts of Deglaciation on the Water Resources of the Himalaya, Centre for Ecology and Hydrology, Oxfordshire, U.K.

Thayyen, R., and J. Gergan. 2009. "Role of Glaciers in Watershed Hydrology: 'Himalayan Catchment' Perspective." *The Cryosphere Discussions* 3: 443–76.

Thornthwaite, C. 1953. "Topoclimatology." Paper presented at Toronto Meteorological Conference, "Symposium on Microclimatology and Micrometeorology," Toronto, Ontario, Canada, September 15.

Wagnon, P., A. Linda, Y. Arnaud, R. Kumar, P. Sharma, C. Vincent, J. G. Pottakkal, E. Berthier, A. Ramanathan, S. I. Hasnain, and P. Chevallier. 2007. "Four Years of Mass Balance on Chhota Shigri Glacier, Himachal Pradesh, India, a New Benchmark Glacier in the Western Himalaya." *Journal of Glaciology* 53 (183): 603–11.

Wake, C. 1988. "Snow Accumulation Studies in the Central Karakoram, Pakistan." *Proceedings of the 44th Eastern Snow Conference*, 19–33. Fredericton, New Brunswick, Canada, June 3–4, 1987.

———. 1989. "Glaciochemical Investigations as a Tool to Determine the Spatial Variation of Snow Accumulation in the Central Karakoram, Northern Pakistan." In *Symposium on Glacier Mapping and Surveying, University of Iceland, Reykjavik, August 26–29, 1985*, edited by E. L. Richardson, 13: 279–84.

World Wildlife Fund. 2005. "An Overview of Glaciers, Glacier Retreat, and Subsequent Impacts in Nepal, India, and China." WWF Nepal Program, Kathmandu.

CHAPTER 4

Future Climate Scenarios for the Indus Basin

Key Messages

- Historical trends show (statistically significant) increasing temperatures and annual precipitation over the last century over the entire country of Pakistan.
- Each river in the Indus has its own hydrologic regime and timing depending on the mix of snow, ice, and monsoon contributions.
- Historically, the annual flows into Tarbela reservoir have been decreasing over time.
- The general findings from a wide range of general circulation model (GCM) outputs show agreement among models regarding continued increases in temperature into the future. Increases are estimated to be at worst close to 3°C warmer by the 2050s. These models are likely more reliable for the irrigated plains than the mountainous upper basin.
- There is not agreement among models regarding changes in precipitation (both in magnitude and direction) because standard errors are large.
- However, there is some indication of a general trend in increased precipitation during the summer and a decrease during the winter. These changes appear to be more pronounced in the southern parts of the country.
- Using the snow and ice hydrology model developed in the previous chapter and a wide range of climate futures, the postulated impact of climate change on inter-annual flow variations is generally comparable with the current inter-annual variations.
- The primary impact on the Upper Indus Basin (UIB) of all but the most extreme climate change scenarios could be a shift in the timing of peak runoff and not a major change in annual volume.
- The review of GCM outputs supports the subsequent modeling approach where ranges of climate risks are informed by GCM outputs but not driven directly by them.

Floods and droughts of the past decade have increased concerns about climate change in the Indus Basin. Analysis of climate variability and change has advanced considerably since the last assessment of potential climate impacts in the Indus Basin 20 years ago (Wescoat and Leichenko 1992). The most comprehensive assessments of climate change risks in Pakistan to date are from the Global Change Impact Study Centre (GCISC) (Ali, Hasson and Khan 2009; Faisal et al. 2009; Iqbal et al. 2009a, 2009b, 2009c; Islam et al. 2009a, 2009b, 2009c; Saeed et al. 2009a, 2009b, 2009c; Saeed, Sheikh, and Khan 2009d; Sheikh et al. 2009; Syed et al. 2009). This chapter will review this work. The climate scenarios described here will be used in subsequent modeling chapters. It is important to underscore that for modeling purposes the Upper and Lower Indus Basin require different hydroclimatic approaches, thus they are treated separately in the following sections.

Overview of Historical Patterns and Trends

Pakistan experiences some of the hottest and driest conditions in the South Asia region, with the exception of the UIB where cool, moist conditions prevail. The upper basin and northern plains lie on the western edge of the monsoon and have comparatively high winter precipitation. The temperature trends for the country as a whole, using Climatic Research Unit (CRU) gridded data, indicate an overall pattern of warming (+0.6°C) over the past century (Sheikh et al. 2009) (figure 4.1). This trend is significant at the 99 percent level. These temperature trends do not display a consistent regional pattern. Annual warming has occurred over the past half-century in the Upper Indus, Punjab plains, and the Balochistan plateau, while some cooling has occurred in the lower parts of the Indus. Examined by season, the patterns are even more complex, as cooler monsoons and hotter summers (April–May) are observed over most of the basin. These variations offset one another in annual temperature trends, which are generally less than 1°C, except in Balochistan and the Western Highlands. Overall seasonal warming and

Figure 4.1 CRU Mean Temperature Data for Pakistan over the 20th Century

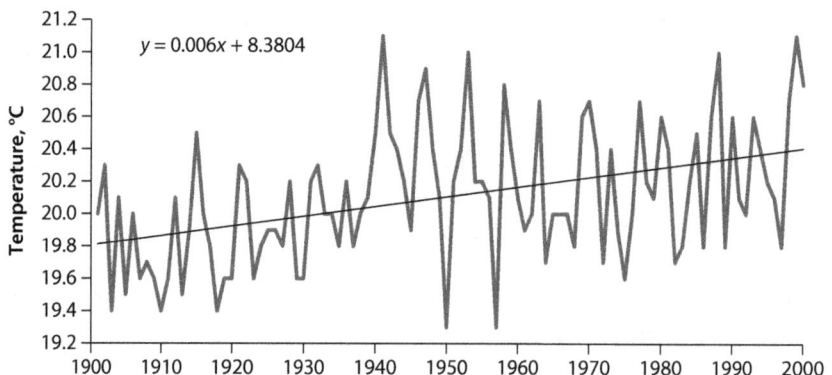

$y = 0.006x + 8.3804$

Source: Sheikh et al. 2009.

cooling trends are somewhat higher, with changes ranging up to 1°C. These patterns of mean temperature are further complicated by different trends in minimum and maximum temperatures in which maximum temperatures have increased over the Upper Indus and decreased over the irrigated basin, while minimum temperatures have decreased over the Upper Indus and Sindh but increased over Punjab and the coastal belt.

Precipitation trends over the country have also increased significantly over the past century. Figure 4.2 indicates a century-long increase of 25 percent, or 63 mm, over the country during the 20th century. This trend is significant at the 99 percent level. Precipitation patterns across provinces and within the year are less clear.

This overall increasing trend in precipitation is apparent over most of the regions in the country. Table 4.1 shows the calculated annual and seasonal average precipitation (1951–2000) across the different regions. Increasing

Figure 4.2 CRU Precipitation Data over Pakistan over the 20th Century

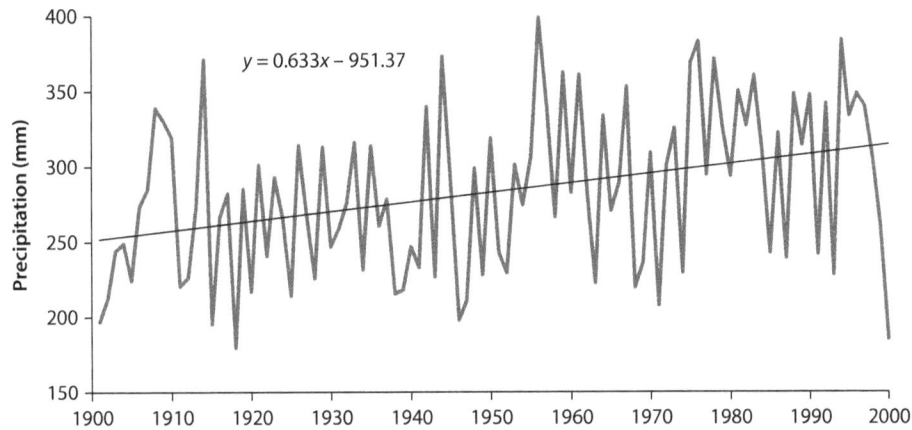

$y = 0.633x - 951.37$

Source: Sheikh et al. 2009.

Table 4.1 Annual and Seasonal Average Precipitation by Zone, 1951–2000
millimeters

Region	Annual average	Monsoon average	Winter average	Apr–May average	Oct–Nov average
Greater Himalayas (winter dominated)	436.3	99.7	185.1	116.6	36.5
Sub-mountain region (monsoon dominated)	1272.9	710.4	352.2	146.1	68.2
Western Highlands	571.1	238.6	201.5	97.8	34.5
Central and Southern Punjab	286.9	189.1	54.7	32.1	10.8
Lower Indus Plains	148.7	120.4	15.1	6.3	5.0
Balochistan Plateau (Northern)	246.0	112.5	92.2	32.2	9.6
Balochistan Plateau (Western)	74.6	13.4	50.5	8.1	3.1
Coastal Belt	155.7	89.3	55.9	4.9	5.9

Source: Sheikh et al. 2009.

Table 4.2 Precipitation Trends, 1951–2000

millimeters

Region	Annual	Monsoon (Jun–Sep)	Winter (Dec–Mar)
Greater Himalayas (winter dominated)	0.49	1.73	–0.04
Sub-mountain region (monsoon dominated)	0.30	0.38	0.53
Western Highlands	–0.02	0.22	0.00
Central and Southern Punjab	0.63	0.57	0.99
Lower Indus Plains	0.22	0.45	–0.27
Balochistan Plateau (Northern)	1.19	1.16	1.14
Balochistan Plateau (Western)	0.10	–0.20	–0.40
Coastal Belt	–0.82	–1.34	0.00

Source: Sheikh et al. 2009.

Table 4.3 Annual Stream Inflow, 1961–2010

million acre-feet

River	Mean	Standard deviation
Indus	60.30	7.37
Chenab	25.48	4.03
Jhelum	22.08	4.70
Kabul	15.93	3.89
Swat	4.69	1.06
Ravi	4.13	2.83
Sutlej	2.59	2.99
Soan	1.07	0.47
Harro	0.73	0.54
Total	135.85	16.05

precipitation trends are strongest over the Upper Indus Punjab and Balochistan (table 4.2), and weaker over the Western Highlands and Coastal Belt. These historical patterns and trends in temperature and precipitation indicate some of the concerns that are arising in Pakistan over increased hydroclimatic risks.

Data from nine river stations (1961–2010) that contribute to the Indus Basin were analyzed: Indus, Chenab, Jhelum, Kabul, Swat, Ravi, Sutlej, Soan, and Harro. Table 4.3 shows the summary statistics; histograms of these flow records are shown in figures 4.3 and 4.4. The Indus mainstream flow varies from 45 to 80 million acre-feet (MAF), and for the Chenab and Jhelum, flow varies from 15 to 35 MAF. The Kabul River, the major surface water supply for North-West Frontier Province (NWFP), has inflow variations from 10 to 30 MAF. All other rivers show annual inflow values less than 10 MAF. Note that the flows from the Ravi and Sutlej (which originate in India) are governed by the Indus Water Treaty with India. Figure 4.5 shows the frequency analysis of these nine tributaries. The total 10 percent exceedance probability for all rivers is 210 MAF and the total 90 percent exceedance probability is 101 MAF. These rivers also show a strong seasonal behavior with most of the flow dominating during the June-September months (figure 4.6).

The time series of the Indus, Chenab, and Jhelum rivers demonstrates the relative stability of these rivers. A simple comparison of the coefficient of variation (CV) of the inter-annual flow of the Indus River with other major rivers in the world (table 4.4) shows the Indus at the lower end: the CV of the Indus is 13 percent, significantly lower than the world average of 49 percent. In comparison, the Ganges, which like the Indus arises from headwaters in the Himalaya, has almost twice the variability, with a CV of 27 percent. This is in

Figure 4.3 Histogram of Annual Indus Inflow

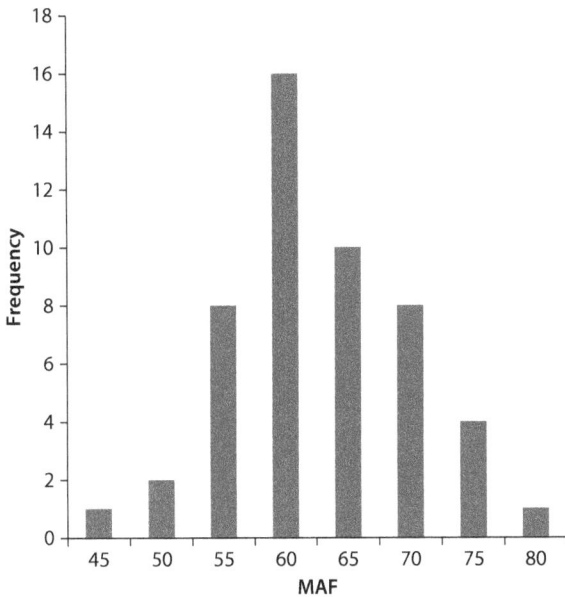

Note: MAF = million acre-feet.

Figure 4.4 Histogram of Annual Chenab, Jhelum, Kabul, Ravi, Sutlej, Swat, Soan, and Harro Inflows

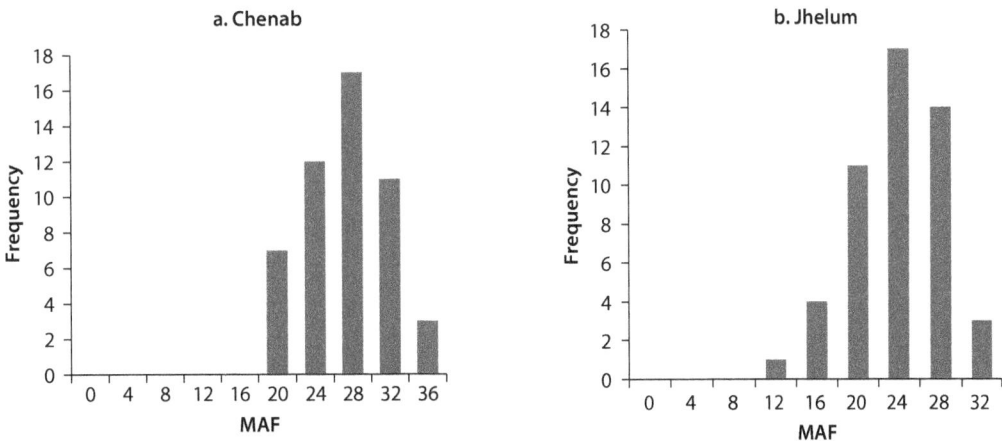

figure continues next page

The Indus Basin of Pakistan • http://dx.doi.org/10.1596/978-0-8213-9874-6

Figure 4.4 Histogram of Annual Chenab, Jhelum, Kabul, Ravi, Sutlej, Swat, Soan, and Harro Inflows (continued)

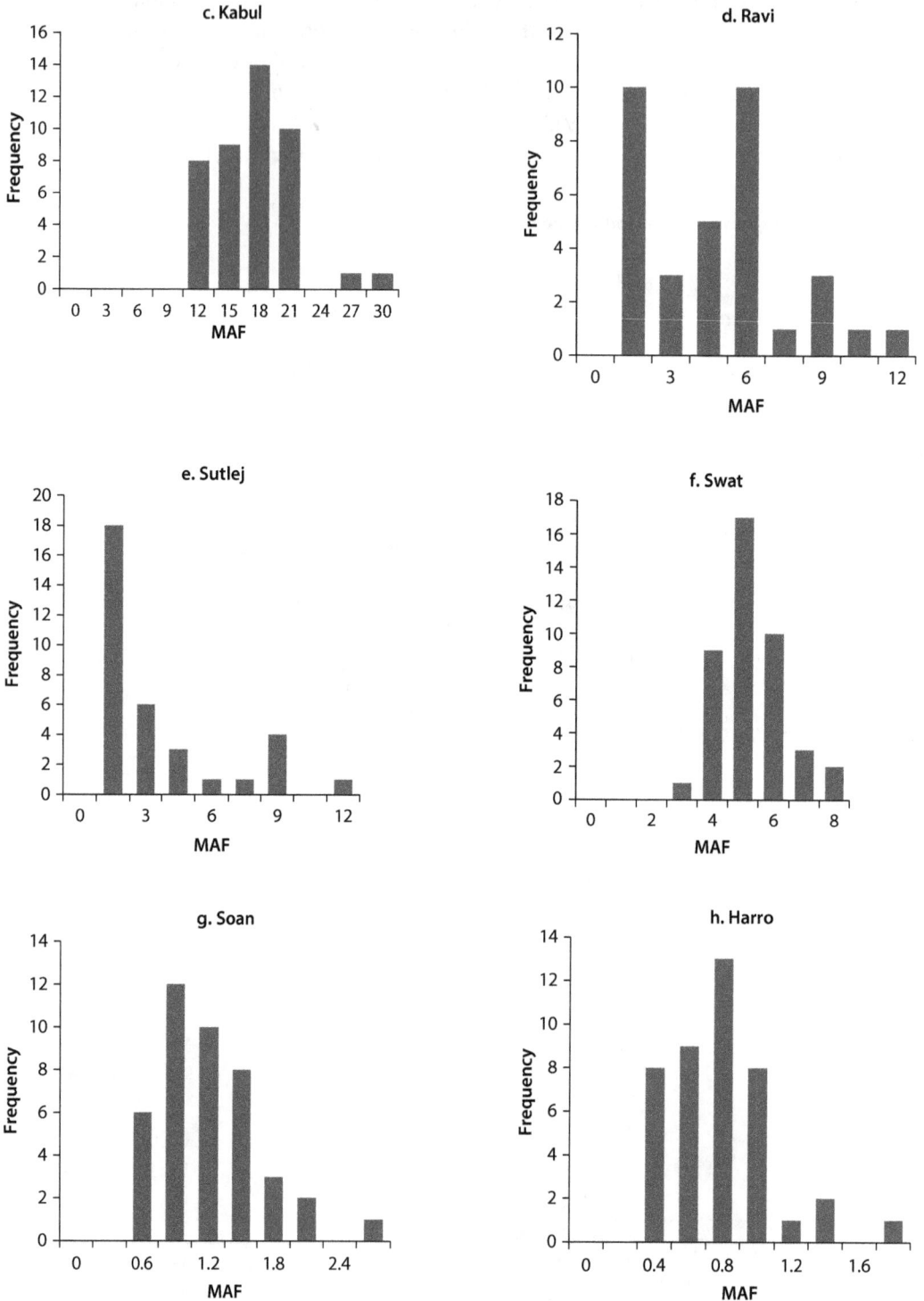

Note: MAF = million acre-feet.

Figure 4.5 Frequency Analysis of Major Tributaries in the Indus Basin

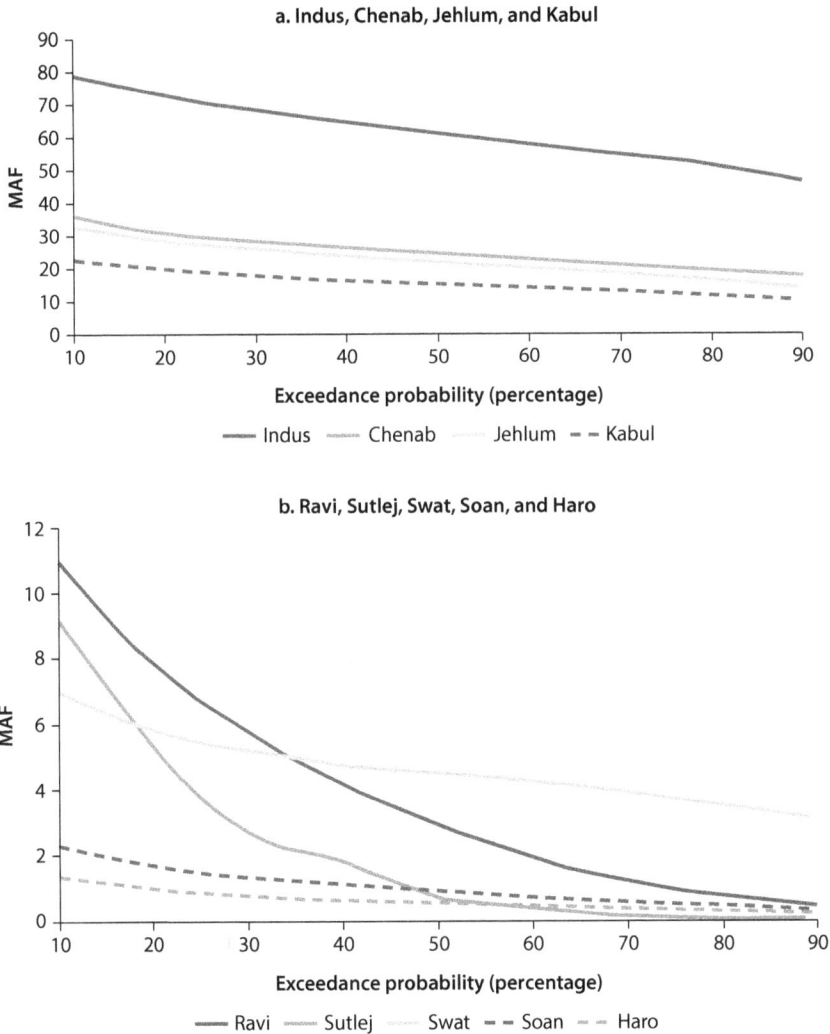

a. Indus, Chenab, Jehlum, and Kabul

— Indus ⋯⋯ Chenab   Jehlum − − Kabul

b. Ravi, Sutlej, Swat, Soan, and Haro

— Ravi ⋯⋯ Sutlej   Swat − − Soan − − Haro

Note: MAF = million acre-feet.

part due to the moderating impact that snow and ice play in the headwaters of the system (as described in chapter 3). However, the impact of climate change on flow variability is still unknown for this region.

Annual historical inflows on the Indus have been declining (significant at 95 percent) over the period of record (figure 4.7). This is contrary to the general idea that an upward trend in discharge would be associated with increasing temperature (figure 4.1), precipitation (figure 4.2) and anticipated increasing melt waters (as discussed in Archer et al. 2010).

Intra-annually, there is some evidence that suggests a slight shift in the hydrograph toward earlier melting and inflows into the Indus. The data show that 21.7 percent of the total distribution flows into Tarbela were in June

Figure 4.6 Average Monthly Inflow in IBMR from Nine Tributaries

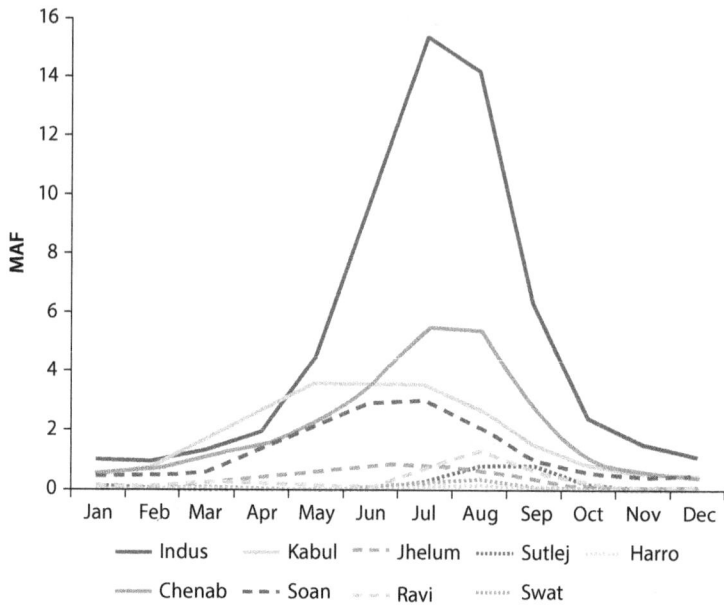

Note: MAF = million acre-feet.

Table 4.4 Coefficient of Variation for Major World Rivers

River	CV (percentage)	Source
Indus	13.0	This study
Amazon	27.0	Villar et al. 2009
Congo	0.3	Global Runoff Data Centre (GRDC)
Ganges	27.0	Mirza et al. 2001
Murray-Darling	60.0	Simpson et al. 1993
Mississippi	21.0	GRDC
Orinoco	14.0	Marengo 1995
Yellow	26.0	Miao and Ni 2009
World average	49.0	Dettinger and Diaz 2000

during the most recent decade in comparison to the earlier period of 40 years, in which only 18 percent of the total flows were in June. Moreover, the Chenab shows a slight increasing trend over time during the *rabi* period, October to March, (statistically significant at 95 percent) and the Indus shows a slightly decreasing trend over time during the *kharif* period, April to September (figure 4.8). An examination of monthly trends showed no clear trends except for a slight positive trend in May on the Chenab and a slight decrease in June on the Jhelum. These types of shifts will be tested in the model runs in later chapters.

Figure 4.7 Indus Inflows, 1937–2011

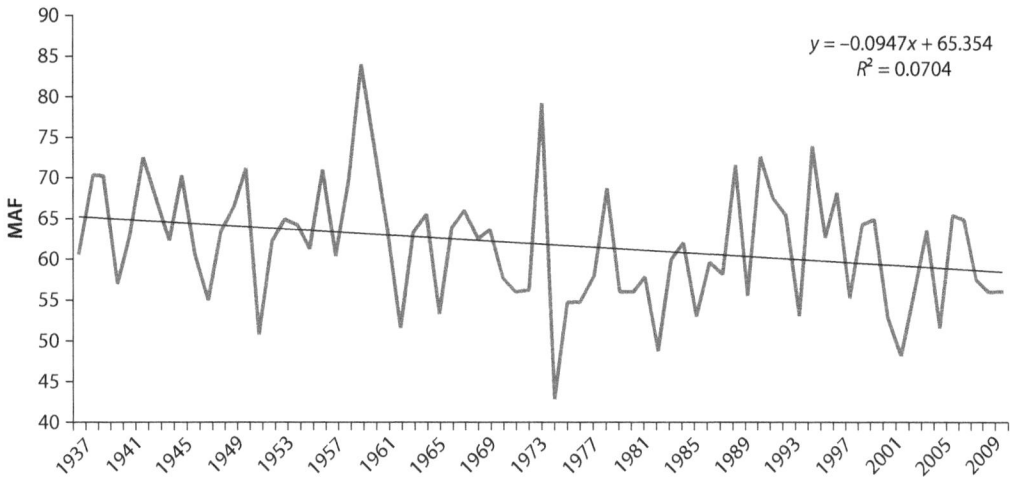

$$y = -0.0947x + 65.354$$
$$R^2 = 0.0704$$

Note: MAF = million acre-feet.

Figure 4.8 Time Series of Flows on the Indus, Jhelum, and Chenab, 1922–2009

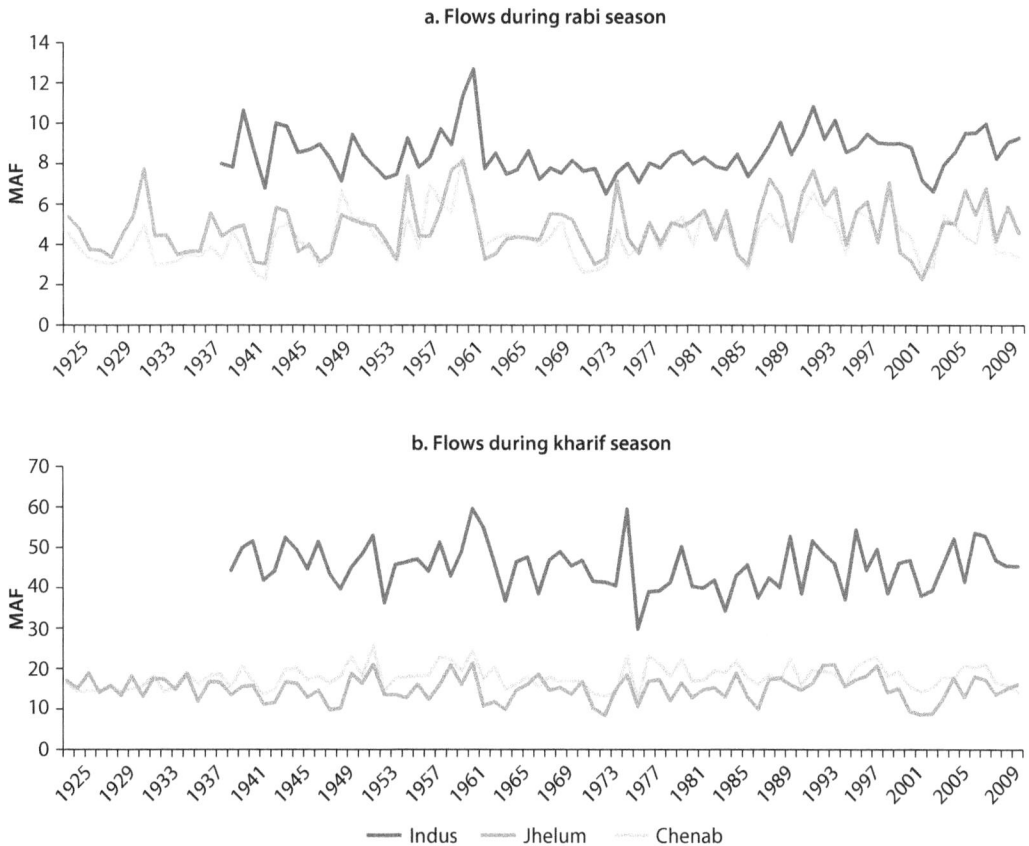

a. Flows during rabi season

b. Flows during kharif season

Indus ——— Jhelum ——— Chenab

Note: MAF = million acre-feet.

Future Climate Change Risks

The UIB and Indus Basin Irrigation System (IBIS) require different approaches to construct climate scenarios and develop the models undertaken in this report. The climates of the two areas are fundamentally different and the geographic scope of the subsequent models is limited to the IBIS region. The UIB has complex terrain and a precipitation regime dominated by westerly waves that generate winter snowfall. By comparison, the IBIS is a summer monsoon-dominated, gently sloping, highly constructed basin that begins below the rim stations. The GCISC analysis of climate change scenarios involves GCM outputs on a 1.0 degree grid, which also enabled comparison between GCM baseline runs and with CRU historical climate patterns.[1] GCMs have been used to project scenarios of climate change under different trajectories of economic development and greenhouse gas emissions.

The GCISC-RR-03 research report (Islam et al. 2009b) drew upon the work of the *IPCC (Intergovernmental Panel on Climate Change) Fourth Assessment Report (AR4)* (IPCC 2007), completed in 2007. GCISC took this opportunity to analyze 17 new GCM model outputs for the AR4 scenarios, individually and in ensemble runs (17 and 13 models were used for the A1B[2] and A2 emissions scenarios respectively. The A2 emissions scenario represents one of the higher emissions scenarios of the future. The GCISC included in the output the mean monthly temperature and precipitation projections for the 2020s, 2050s, and 2080s. Although the 2080 projections are used by climatologists, they are unrealistic as far accuracy of predictions of future water and agricultural systems. There is no credible way to anticipate linkages among these systems out to the 2080s, and even the 2050s are likely to have numerous unforeseen surprises. Only the GCMs that matched the historical normals (1961–90) well (here defined as less than 2°C difference and less than 20 percent difference in precipitation) were examined in the GCISC analysis.

Future Climate in the Indus Basin Irrigation System

These GCISC modeling results (figure 4.9) show that by the 2020s the temperature is expected to rise by about 2°C in northern Pakistan, 1.5°C in the central parts of the country, and 1°C in the coastal areas. Temperatures will continue to increase into the 2050s and 2080s. As for precipitation, the changes in Pakistan are not conclusive, even out to 2080s. These impacts are even less for the A1B and B2 emissions scenarios. This highlights the difficulty in making both estimates of magnitude and direction for precipitation.

Focusing more on comparisons between the northern parts of Pakistan (that is, UIB) and the southern parts (that is, IBIS), simulations of 17 GCMs, the ensemble values (and standard errors) are shown in tables 4.5 and 4.6. For both the A2 and A1B scenarios, temperatures are likely to be near 4°C warmer by 2080. Moreover, under these GCISC model runs, it is difficult to say with certainty how precipitation will change, because the uncertainties are large in all cases.

Moreover, it is important to consider seasonal changes as well (tables 4.7 and 4.8). In the northern region this means focusing on winter precipitation. In most GCM projections, winter temperatures increase only slightly more than summer temperatures. Seasonal precipitation differences are significant, but they vary so much in sign and magnitude as to defy generalization across GCM models, and we therefore concentrate on sensitivity analysis (chapter 6) (Islam et al. 2009a, 2009b, 2009c). In light of these seasonal results, the study team did not analyze more disaggregated GCM monthly output.

Figure 4.9 Ensemble Change of Temperature and Precipitation for the A2 Scenario

a. Mean annual ensemble change of temperature (°C)

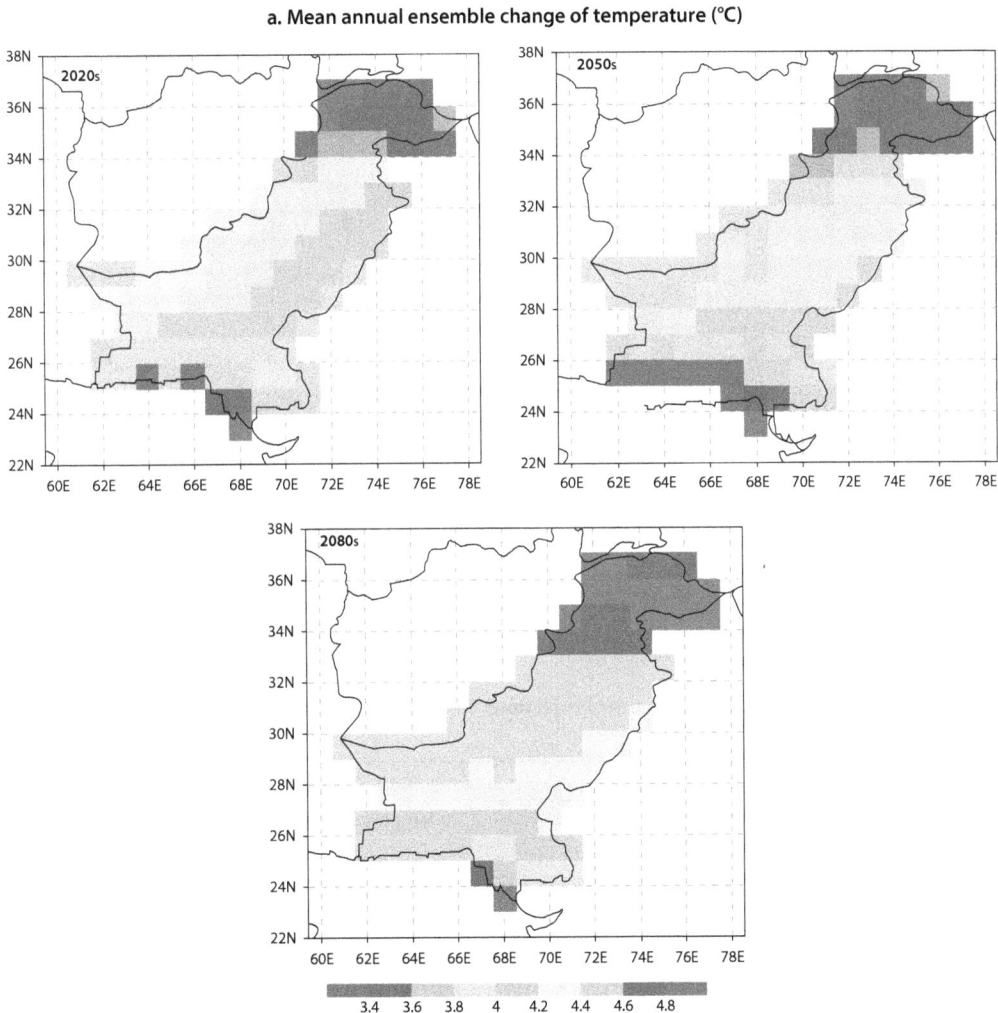

figure continues next page

Figure 4.9　Ensemble Change of Temperature and Precipitation for the A2 Scenario *(continued)*

b. Mean annual ensemble change of precipitation (percentage)

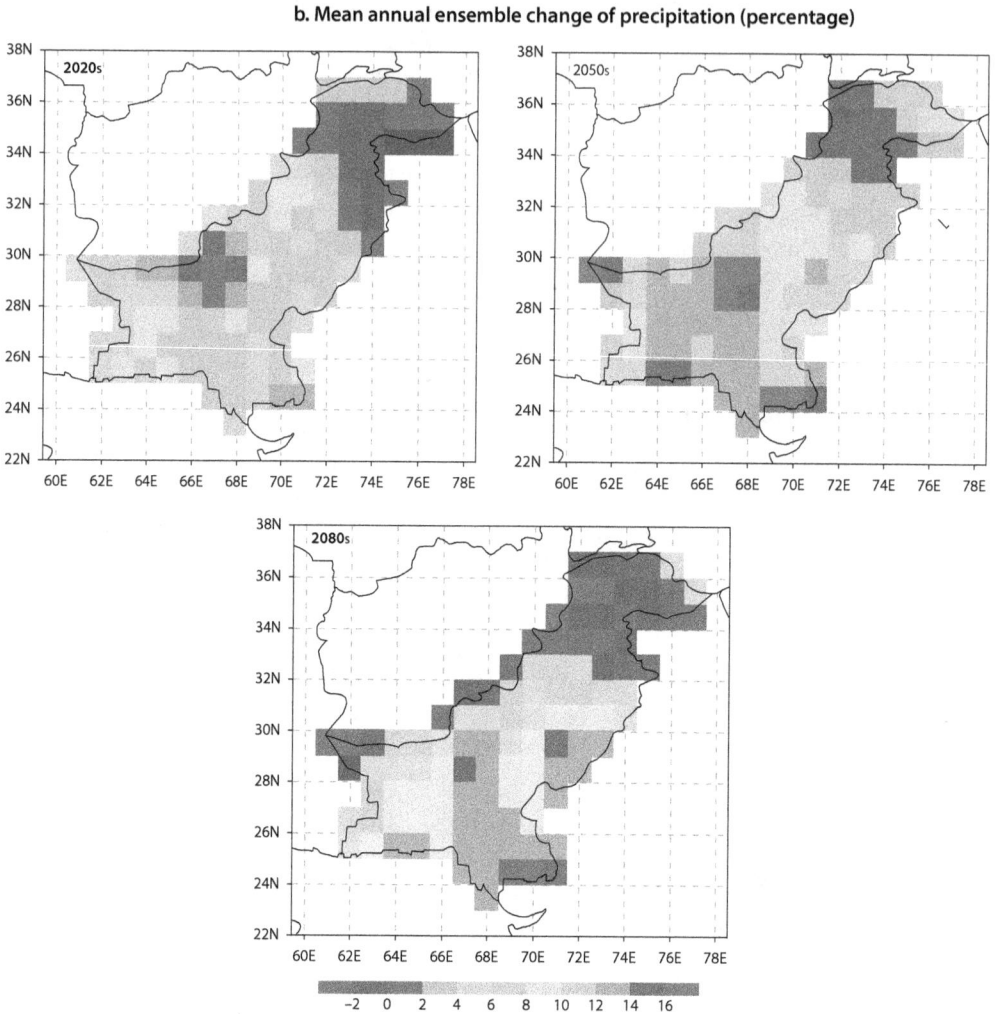

Source: © Global Change Impact Study Centre. Reproduced, with permission, from Islam et al. 2009b; further permission required for reuse.

Table 4.5　Ensemble Mean of Climate Change Projections Based on IPCC-AR4 Using 17 GCMs and the A2 Special Report on Emissions Scenarios

Area	2020s	2050s	2080s
		Temperature change (°C)	
Northern Pakistan	1.4 ± 0.1	2.7 ± 0.2	4.7 ± 0.2
Southern Pakistan	1.3 ± 0.1	2.4 ± 0.1	4.2 ± 0.2
		Precipitation change (%)	
Northern Pakistan	2.2 ± 2.3	3.6 ± 3.2	1.1 ± 4.0
Southern Pakistan	3.1 ± 5.1	6.4 ± 7.5	4.3 ± 9.4

Source: GCISC.

Note: IPCC-AR4 = Intergovernmental Panel on Climate Change Fourth Assessment Report, GCM = general circulation model.

Table 4.6 Ensemble Mean of Climate Change Projections Based on IPCC-AR4 Using 17 GCMs and the A1B Special Report on Emissions Scenarios

Area	2020s	2050s	2080s
	Temperature change (°C)		
Northern Pakistan	1.6 ± 0.1	3.0 ± 0.2	4.1 ± 0.2
Southern Pakistan	1.4 ± 0.1	2.6 ± 0.1	3.7 ± 0.2
	Precipitation change (%)		
Northern Pakistan	−0.7 ± 1.5	−1.8 ± 2.2	−0.7 ± 3.1
Southern Pakistan	−3.2 ± 4.3	−0.3 ± 5.5	−0.9 ± 7.9

Source: GCISC.
Note: IPCC-AR4 = Intergovernmental Panel on Climate Change Fourth Assessment Report, GCM = general circulation model.

Table 4.7 Projected Summer (JJAS) Changes Using 17 GCMs

Scenario/area	A2		A1B	
	2020s	2050s	2020s	2050s
	Temperature change (°C)			
Northern Pakistan	1.3 ± 0.1	2.6 ± 0.2	1.5 ± 0.1	2.9 ± 0.2
Southern Pakistan	1.1 ± 0.1	2.2 ± 0.2	1.2 ± 0.1	2.4 ± 0.2
	Precipitation change (%)			
Northern Pakistan	5.5 ± 3.7	7.6 ± 6.5	1.3 ± 3.0	1.8 ± 4.7
Southern Pakistan	12.5 ± 9.8	42.2 ± 27.0	11.2 ± 11.0	24.1 ± 18.1

Source: GCISC.
Note: JJAS = June, July, August, and September, GCM = general circulation model.

Table 4.8 Projected Winter (ONDJ) Changes Using 17 GCMs

Scenario/area	A2		A1B	
	2020s	2050s	2020s	2050s
	Temperature change (°C)			
Northern Pakistan	1.5 ± 0.1	2.8 ± 0.2	1.7 ± 0.1	3.0 ± 0.2
Southern Pakistan	1.4 ± 0.1	2.6 ± 0.1	1.6 ± 0.1	2.8 ± 0.1
	Precipitation change (%)			
Northern Pakistan	−0.7 ± 2.3	0.7 ± 3.2	−2.6 ± 1.9	−4.7 ± 2.6
Southern Pakistan	−7.5 ± 6.1	−12.9 ± 6.8	−16.1 ± 4.7	−9.9 ± 7.3

Source: GCISC.
Note: ONDJ = October, November, December, and January, GCM = general circulation model.

The temperature increases in both summer and winter are higher in northern Pakistan than in southern Pakistan. Moreover, temperature increases tend to be on average higher during the winter than the summer. General trends are more difficult to surmise with precipitation because the standard errors are large. There is some indication, however, of a general trend in increased precipitation during the summer and a decrease during the winter. The changes appear to be more pronounced in the southern parts of the country.

Future Climate in the Upper Indus Basin

Though debate is ongoing as to the scientific soundness of using GCMs to make predictions in high topography areas, the GCISC-RR-03 report (Islam et al. 2009b.) provides predicted changes in temperature and precipitation in the UIB (table 4.9).

There is agreement among models regarding increases in temperature. But, the direction of precipitation is unclear because the standard errors are large. Moreover, a key question is how these changes might affect river discharge and inflows into the Indus Basin. Using the UIB model developed in chapter 3 and assuming a much larger change than is being predicted by the 17 GCMs above, scenarios can be generated of different inflows into the Tarbela Dam on the Indus main stem (table 4.10).

These future projections indicate that the simple physical ablation model yields inflows into the Indus main stem ranging from 45.4 to 73.8 MAF, or from 78 to 126 percent of the mean historical discharge of the Indus at Tarbela.

Table 4.9 Ensemble Mean of Climate Change Projections for the Upper Indus Basin Based on IPCC-AR4 Using 17 GCMs and the A2 Special Report on Emissions Scenarios

	2020s	2050s	2080s
		A2 Scenario	
Temperature changes	1.48 ± 0.12	2.84 ± 0.17	4.84 ± 0.24
Precipitation changes	0.60 ± 1.55	2.47 ± 1.82	1.84 ± 2.36
		A1B	
Temperature changes	1.63 ± 0.11	3.10 ± 0.21	4.36 ± 0.31
Precipitation changes	−0.15 ± 1.04	0.28 ± 1.60	2.08 ± 2.26

Source: GCISC.

Table 4.10 First-Order Effects of Temperature and Precipitation Changes on Discharge into the Indus Main Stem

Temperature changes (°C)	Precipitation changes, MAF (percentage of baseline)				
	−20%	−10%	No change	+10%	+20%
+0.5	45.4 (78)	50.3 (86)	55.1 (94)	60 (103)	64.9 (111)
+1.5	46.2 (79)	51.1 (88)	55.9 (96)	60.8 (104)	65.7 (113)
+2.0	46.2 (79)	51.1 (88)	55.9 (96)	60.8 (104)	65.7 (113)
+3.0	48.6 (83)	53.5 (92)	58.4 (100)	63.2 (108)	68.1 (117)
+4.0	52.7 (90)	57.6 (99)	62.4 (107)	67.3 (115)	72.2 (124)
+4.5	54.3 (93)	59.2 (101)	64 (110)	68.9 (118)	73.8 (126)

Note: Baseline temperature and precipitation gives an average of 48.7 million acre-feet (MAF) of snowmelt and 9.7 MAF of ice melt, for a total baseline of 58.4 MAF. Percentage change in precipitation is assumed to be directly proportional to changes in snowmelt contributions to runoff. For increases in T, the ablation gradient concept was used, as described in chapter 3. As the temperature increases, the firn line moves upward on the glacier, producing an increase in the surface area of the ablating portion of the glacier (the "ablation facies"). For example, for a 0.5 degree increase in temperature and a 10 percent increase precipitation, it is calculated that the snowmelt contribution would increase to 53.5 MAF and, based on the earlier analysis, the ice melt contribution would decrease to 6.5 MAF. However, the total runoff would increase to 60 MAF (103 percent above the baseline).

Figure 4.10 Future Indus Inflow Histogram Using the UIB Model

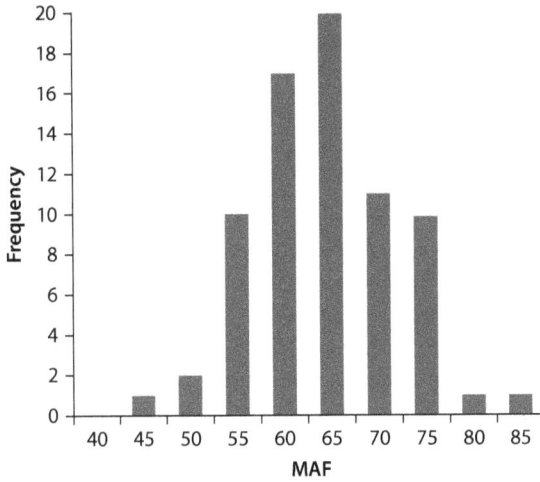

Note: MAF = million acre-feet, UIB = Upper Indus Basin.

Interestingly, the histogram of melt water estimates (figure 4.10) are quite similar to the historical discharge volumes (see figure 4.3, page 81).

Thus, based on the analyses of this study, it is estimated that the present inter-annual variations in stream flow from the tributaries of the UIB are generally comparable to the postulated impacts of the climate change scenarios currently being applied to the mountains of South Asia. So it can be concluded that the primary impact of all but the most extreme climate change scenarios will be a shift in the timing of peak runoff, and not a major change in annual volume.

Notes

1. Each of the GCM models analyzed uses a different grid. GCM grid sizes range from 2.8 × 2.8 to 5.6 × 5.6 degrees (GCISC-RR-02 [Faisal et al. 2009] and RR-04 [Saeed, Sheikh, and Khan 2009d, 4]). GCISC worked with the University of Trieste to interpolate model output values for all grid cells onto the same 0.5-degree grid to compare the baseline runs.

2. The A1 scenario family describes a future world of very rapid economic growth, global population that peaks in mid-century and declines thereafter, and the rapid introduction of new and more efficient technologies. The A1B scenario is a "balance" scenario across all sources. Where balanced is defined as not relying too heavily on one particular energy source, on the assumption that similar improvement rates apply to all energy supply and end-use technologies. The A2 scenario family describes a very heterogeneous world. The underlying theme is self-reliance and preservation of local identities. Fertility patterns across regions converge very slowly, which results in continuously increasing population.

References

Ali, G., S. Hasson, and A. M. Khan. 2009. *Climate Change: Implications and Adaptation of Water Resources in Pakistan.* Research Report GCISC-RR-13, Global Change Impact Study Centre, Islamabad.

Archer, D. R., N. Forsythe, H. J. Fowler, and S. M. Shah. 2010. "Sustainability of Water Resources Management in the Indus Basin under Changing Climatic and Socio Economic Conditions." *Hydrology and Earth System Sciences* 14: 1669–80.

Dettinger, M. D. and H. F. Diaz. 2000. "Global Characteristics of Stream Flow Seasonality and Variability." *Journal of Hydrometeorology* 1: 289–310.

Faisal, S., S. Syed, S. Islam, N. Rehman, M. M. Sheikh, and A. M. Khan. 2009. *Climate Change Scenarios for Pakistan and Some South Asian Countries for SRES A2 and B2 Scenarios Based on Six Different GCMs Used in IPCC-TAR.* Research Report GCISC-RR-02, Global Change Impact Studies Centre, Islamabad.

IPCC (Intergovernmental Panel on Climate Change). 2007. *Synthesis Report: Contribution of Working Groups I, II and III to the Fourth Assessment Report of the Intergovernmental Panel on Climate Change,* edited by R. K. Pachauri and A. Reisinger. IPCC, Geneva, Switzerland.

Iqbal, M. M., M. A. Goheer, S. A. Noor, H. Sultana, K. M. Salik, and A. M. Khan. 2009a. *Climate Change and Agriculture in Pakistan: Adaptation Strategies to Cope with Negative Impacts.* Research Report GCISC-RR-16, Global Change Impact Studies Centre, Islamabad.

———. 2009b. *Climate Change and Rice Production in Pakistan: Calibration, Validation and Application of CERES-Rice Model.* Research Report GCISC-RR-15, Global Change Impact Studies Centre, Islamabad.

———. 2009c. *Climate Change and Wheat Production in Pakistan: Calibration, Validation and Application of CERES-Wheat Model.* Research Report GCISC-RR-14, Global Change Impact Studies Centre, Islamabad.

Islam, S., N. Rehman, M. M. Sheikh, and A. M. Khan. 2009a. *Assessment of Future Changes in Temperature Related Extreme Indices over Pakistan Using Regional Climate Model PRECIS.* Research Report GCISC-RR-05, Global Change Impact Study Centre, Islamabad.

———. 2009b. *Climate Change Projections for Pakistan, Nepal and Bangladesh for SRES A2 and A1B Scenarios Using Outputs of 17 GCMs Used in IPCC-AR4.* Research Report GCISC-RR-03, Global Change Impact Study Centre, Islamabad.

———. 2009c. *High Resolution Climate Change Scenarios over South Asia Region Downscaled by Regional Climate Model PRECIS for IPCC SRES A2 Scenario.* Research Report GCISC-RR-06, Global Change Impact Study Centre, Islamabad.

Marengo, J. A. 1995. "Variations and Change in South American Streamflow." *Climatic Change* 31, 99–117.

Miao, C. Y., and J. R. Ni. 2009. "Variation of Natural Streamflow since 1470 in the Middle Yellow River, China." *International Journal of Environmental Research and Public Health* 6: 2849–64.

Mirza, M. M. Q., R. A. Warrick, N. J. Ericksen, and G. J. Kenny. 2001. "Are Floods Getting Worse in the Ganges, Brahmaputra and Meghna Basins?" *Environmental Hazards* 3 (2): 37–48.

Saeed, F., M. R. Anis, R. Aslam, and A. M. Khan. 2009a. *Comparison of Different Interpolation Methods for Temperature Mapping in Pakistan.* Research Report GCISC-RR-10, Global Change Impact Study Centre, Islamabad.

———. 2009b. *Development of Climate Change Scenarios for Specific Sites Corresponding to Selected GCM Outputs, Using Statistical Downscaling Techniques.* Research Report GCISC-RR-09, Global Change Impact Study Centre, Islamabad.

Saeed, F., S. Jehangir, M. Noaman-ul-Haq, W. Shafeeq, M. Z. Rashmi, G. Ali, and A. M. Khan. 2009c. *Application of UBC and DHSVM Models for Selected Catchments of Indus Basin Pakistan.* Research Report GCISC-RR-11, Global Change Impact Study Centre, Islamabad.

Saeed, S., M. M. Sheikh, and A. M. Khan. 2009d. *Validation of Regional Climate Model PRECIS over South Asia.* Research Report GCISC-RR-04, Global Change Impact Study Centre, Islamabad.

Sheikh, M. M., N. Manzoor, M. Adnan, J. Ashraf, and A. M. Khan. 2009. *Climate Profile and Past Climate Changes in Pakistan.* Research Report GCISC-RR-01, Global Change Impact Study Centre, Islamabad.

Simpson, H. J., M. A. Cane, A. L. Herczeg, S. E. Zebiak, and J. H. Simpson. 1993. "Annual River Discharge in Southeastern Australia Related to El Nino–Southern Oscillation Forecasts of Sea Surface Temperature." *Water Resources Research* 29 (11): 3671–80.

Syed, F. S., S. Mehmood, M. A. Abid, M. M. Sheikh, and A. M. Khan. 2009. *Validation of the Regional Climate Model RegCM3 over South Asia.* Research Report GCISC-RR-07, Global Change Impact Study Centre, Islamabad.

Villar, J. C. E., J. L. Guyot, J. Ronchail, G. Cochonneau, N. Filizola, P. Fraizy, D. Labat, E. de Oliveira, J. J. Ordonez, and P. Vauchel. 2009. "Contrasting Regional Discharge Evolutions in the Amazon Basin (1974–2004)." *Journal of Hydrology* 375 (3–4): 297–311.

Wescoat, J., and R. Leichenko. 1992. "Complex River Basin Management in a Changing Global Climate: Indus River Basin Case Study in Pakistan—A National Modeling Assessment." Collaborative Paper 5, Center for Advanced Decision Support for Water and Environmental Systems, and Civil, Environmental, and Architectural Engineering, University of Colorado, Boulder, CO.

CHAPTER 5

Modeling Water, Climate, Agriculture, and the Economy

This chapter describes the two models used in this integrated modeling framework designed for this study. The first model is a hydro-economic optimization model that takes a variety of inputs (for example, agronomic information, irrigation system data, and water inflows) to generate the optimal crop production across the provinces (subject to a variety of physical and political constraints) in the Indus Basin for every month of the year. This simulation of the hydro-economy of the Indus Basin Irrigation System (IBIS) will respond to changing climate factors and allow the evaluation of different possible investments in terms of their economic contribution. The second model is a computable general equilibrium (CGE) model for the Pakistan macro-economy. This integration helps to better illuminate how changes in climate may impact the macro-economy and different household groups through the agriculture sector. This chapter describes how these two models are integrated and are used to examine investment scenarios in chapter 6.

Section 1: Indus Basin Model Revised (IBMR-2012)

This section describes the basic design and the latest modifications and updates to the Indus Basin Model Revised (IBMR-2012). This model was originally developed in Ahmad, Brooke, and Kutcher (1990) and Ahmad and Kutcher (1992) for the purposes of investment planning for the Water and Power Development Authority (WAPDA). Details of this model can be found in these documents.

Economic Objective

The IBMR primarily covers the provinces of Punjab and Sindh. Only a small percentage of the total command area of the IBIS is in Balochistan and North-West Frontier Province (NWFP). Each province contains one or more agro-climatic zones (ACZs), which are further subdivided based on the cropping pattern, land characteristics, and climatic condition. Twelve ACZs are currently

Figure 5.1 The Zonal Supply-Demand Relationship in IBMR-2012

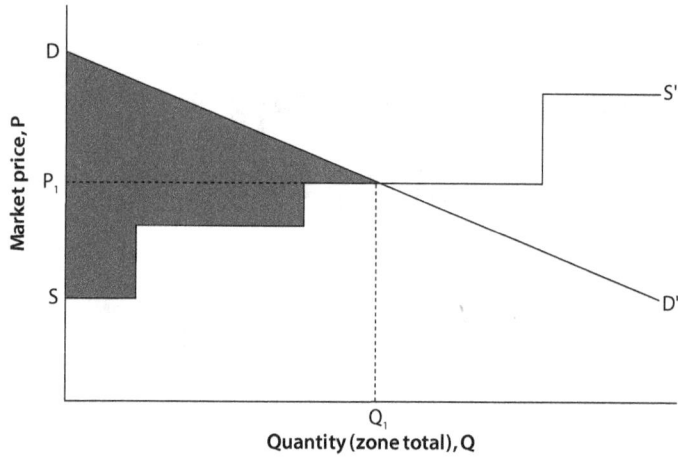

Source: Ahmad, Brooke, and Kutcher 1990.
Note: IBMR-2012 = Indus Basin Model Revised.

used in the IBMR (see descriptions in appendix A). The overall objective function of the IBMR is to maximize the consumer and producer surplus (CPS) for the entire IBIS. Figure 5.1 shows a zonal supply-demand relationship for an individual crop commodity.

The supply function SS' is built from the embedded farm production models in the overall optimization model. The step-wise nature of the supply curve reflects different efficient production technologies, groundwater types, and water stresses (for example, semi-mechanized production using fresh groundwater). The downward sloping linear demand function DD' is constructed using data on the baseline (observed) equilibrium quantity, price, and estimated price elasticity (from Ahmad, Brooke, and Kutcher 1990). The linear format for the demand function is defined as:

$$P = a + bQ$$

Given observed P_0 and Q_0 and elasticity (η)

$$\eta = dQ\,dP \times P/Q$$

We can solve for the slope (b)

$$b = dP/dQ = (P_0/Q_0)/\eta$$

and the intercept (a)

$$a = P_0 - bQ_0$$

The shaded area in figure 5.1 represents the CPS. The objective function of IBMR is to maximize this area. The CPS is a nonlinear function, thus

Figure 5.2 Conceptual Diagram of IBMR

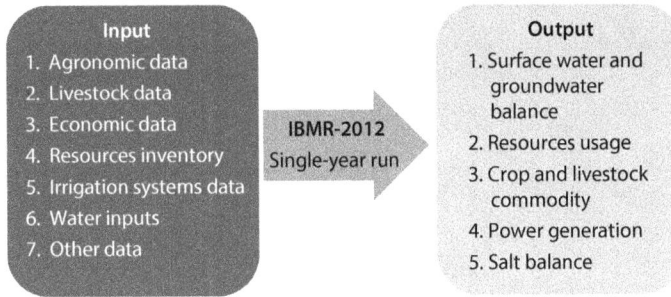

Input		Output
Input 1. Agronomic data 2. Livestock data 3. Economic data 4. Resources inventory 5. Irrigation systems data 6. Water inputs 7. Other data	**IBMR-2012** Single-year run	**Output** 1. Surface water and groundwater balance 2. Resources usage 3. Crop and livestock commodity 4. Power generation 5. Salt balance

Note: IBMR-2012 = Indus Basin Model Revised.

the IBMR uses a piecewise linear programming approach to solve it. Although the model prices may fluctuate between zero and the intercept of the demand curve, this is not likely to happen in reality, thus prices are given upper and lower bounds. It is assumed that outside of these bounds inter-zonal trade will exist. However, the model does not actually simulate trade. The IBMR also does not consider international trade explicitly, but it does account for the prices of international exports and imports and adjusts production accordingly.

Figure 5.2 is a conceptual diagram that explains the modeling process of IBMR. With various input datasets, using month as the modeling time step, the results from IBMR will maximize the single-year CPS and generate outputs such as (1) the monthly surface water and groundwater balance in the Indus Basin, (2) resource usage for each production activity, and (3) monthly production of crop and livestock commodities. Side calculations are used to determine the monthly power generation from reservoirs and the salt balances from the optimized system.

The IBMR has a hierarchal spatial structure for the optimization process. A node-link system is used to represent the entire river and canal network. This system provides surface water supply to each ACZ and simulates the agricultural production and consumption at the ACZ level. In each ACZ, the culturable command area (CCA) is defined based on the existing canal diversions. The hierarchal structure of the IBMR and the zone definitions are summarized in appendix A (see table A.1). The maps of the IBMR ACZ and CCA areas are shown in maps 5.1 and 5.2.

Water Balance

The basic water balance unit in IBMR is the ACZ. The model uses the network theory concepts of nodes and arcs to simulate the flows throughout the system of Indus rivers and link canals. Figure 5.3 shows the complete node-link system map for the Indus River Basin. At each node, a water flow decision is made and the water balance for each month is calculated. The IBMR has 47 nodes that represent reservoirs, inflow stations, barrages, confluences of rivers, and

Map 5.1 Indus River and IBMR Agro-Climatic Zones in Pakistan

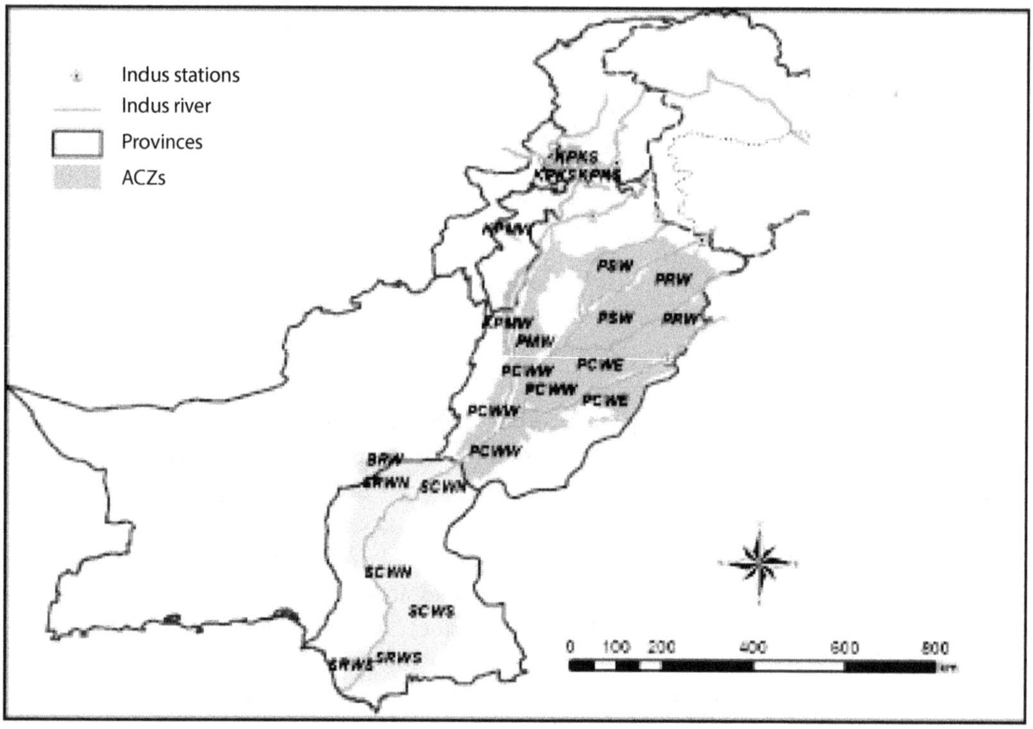

Note: ACZ = agro-climatic zones, IBMR = Indus Basin Model Revised. Details of each ACZ are given appendix A.

the terminus of the Arabian Sea. Forty-nine sinks (terminal nodes) represent diversions to irrigation canals. Finally, 110 arcs represent river reaches and link canals between nodes. Flows along these river reaches are simulated with losses and gains from river bank storage. The IBMR also considers the efficiency of manmade surface channels, such as canals and watercourses for irrigation purposes, which are drawn from river reaches. Figure 5.4 provides a complete sketch of the various losses included in the surface water balance. The stream recharge to groundwater is computed as river seepage and treated as a loss from surface water. The canal water diversion efficiency and watercourse diversion efficiency are considered losses from surface water and as additions to groundwater.

In the IBMR the residual moisture in the root zone is explicitly modeled and represents a potential source of water for crops. That is, the root zone water balance is also the crop water requirement water balance. Thus, crop water needs are met from precipitation, canals, groundwater wells, and the moisture in the root zone (also known as "sub-irrigation"). An evaporation parameter in the model is used to define the sub-irrigation water available to plants. The IBMR assumes that 60 percent of the evaporation from groundwater can be absorbed by crops. Figure 5.5 illustrates this water balance and demonstrates how surface and groundwater interact.

Map 5.2 Indus River and IBMR Canal Command Areas in Pakistan

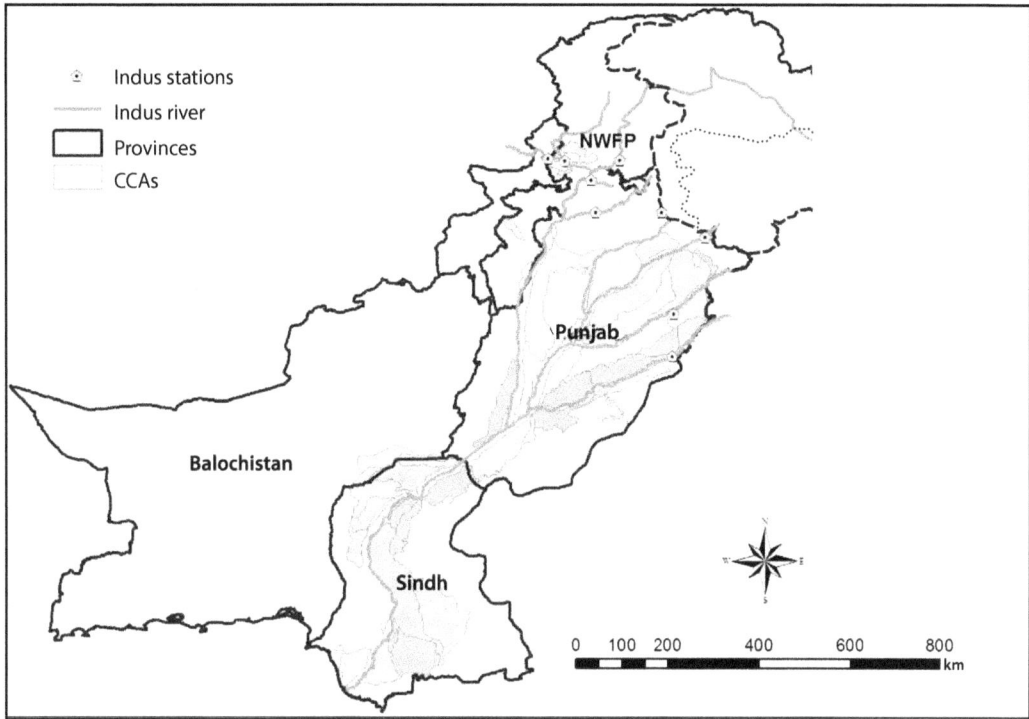

Note: IBMR = Indus Basin Model Revised.

Input Data, Equations, Constraints, and Output Data

The input data for the IBMR has been refined over the years by WAPDA and various researchers (Ahmad, Brooke, and Kutcher 1990). Appendix B provides a discussion of the data used to update the model to a 2008 baseline. The input data of IBMR can be categorized as (1) agronomic and livestock data, (2) economic data, (3) resources inventory, and (4) irrigation systems and water data.

Agronomic and Livestock Data

The required agronomic inputs include crop growing period, labor, crop water needs, fertilizer use, draft power requirements, and crop yield and by-products. There are 14 crops in the current version of the IBMR: basmati rice, irrigated rice, cotton, rabi season fodder, gram, maize, mustard and rapeseed, Kharif season fodder, sugarcane, wheat, orchard, potatoes, onions, and chilies. Information on the water requirements for each crop by month and the types of water application (for example, standard, stressed) is also specified. A different future crop mix is not considered in this model. Livestock data include the labor and the feed requirements of each animal. Conversion factors are used to determine the by-products from these animals: meat and milk. The three types of livestock in the IBMR are bullocks, cows, and buffalo.

Figure 5.3 The Node-Link System Map of IBMR

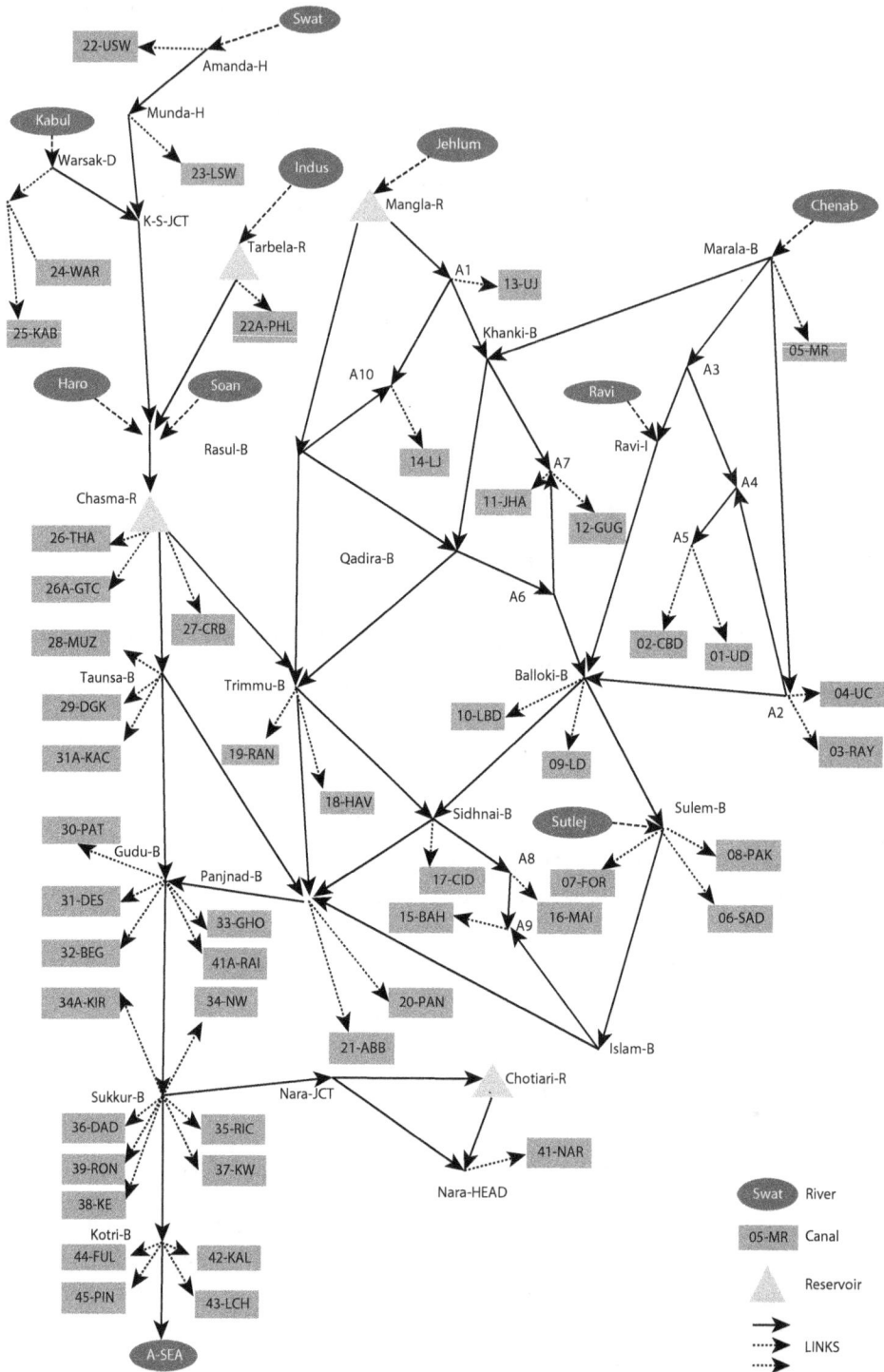

Note: IBMR = Indus Basin Model Revised. Solid line = streamflow. Bold dotted line = stream inflow. Thin dotted line = canal diversions.

Figure 5.4 Surface-Water Balance in IBMR

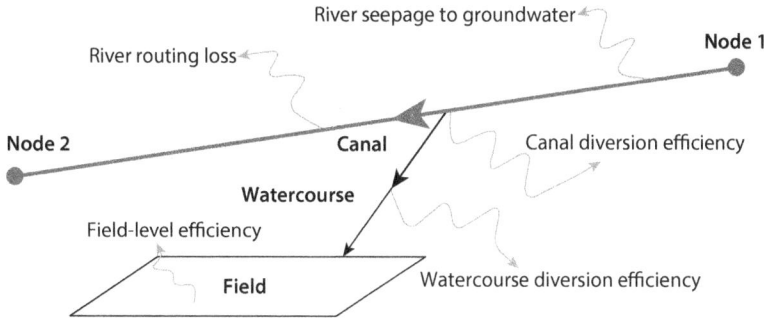

Note: IBMR = Indus Basin Model Revised.

Figure 5.5 Water Balance in IBMR

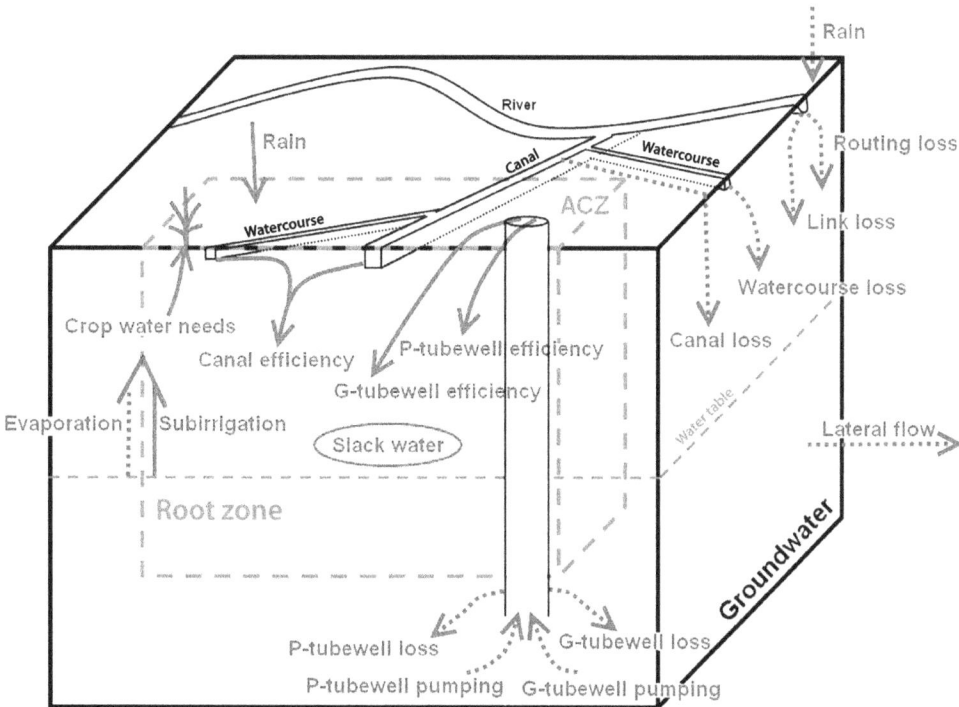

Note: IBMR = Indus Basin Model Revised. The solid lines indicate the root zone water balance components used to supply the crop. The dashed lines represent the groundwater balance components tracked during the simulation runs. All water balance calculations are at the agro-climatic zone (ACZ) scale.

Economic Data

Economic data include estimated demand for each crop and livestock product, market prices for all commodities, fixed cost for meat and orchards, and price elasticities for different commodities. Price elasticities are used from an earlier version of the IBMR. On-farm consumption of grown agriculture crops is included in the model. In the current version of IBMR, consumption crops are identified (basmati, irrigated rice (irri), gram, maize, mustard and rapeseed, sugarcane, wheat, onion, potatoes and chilies), and the on-farm consumption is calculated as proportions of total production based on earlier farm surveys. Depending on the crop, the percentage of total output consumed varies from about 10–50 percent. Once the ACZ production meets the household consumption requirement, the remaining production is available to the market. Therefore, the demand in IBMR is the residual of production less on-farm consumption and is used to fix the quantity axis of the demand function in figure 5.1.

Resources Inventory

Input resources include agricultural workers, tractors, private tubewells, households, animals, and total available irrigated area (cropped land). The farm population is used to compute the labor availability. The IBMR was updated to reflect power requirements (for land preparation) from tractors instead of animals. Almost all power requirements are provided by tractor in the Indus Basin today.

Irrigation Systems and Water Data

When modeling the irrigation system, the basic unit of area is the canal command. All the data availability on these commands must be aggregated to the level of the agricultural model (that is, ACZ). In each ACZ, the cropping pattern and associated technologies are defined and assumed consistent. An ACZ may cut across canal command areas and need not be geographically contiguous; therefore, sub-areas are defined. Fresh and saline groundwater areas are defined and are treated separately; tubewell pumping is allowed in fresh but not in saline areas. The physical characteristics of the canal system are also given, including the culturable command area, canal capacity at the head, canal efficiency, watercourse command efficiencies, and field efficiencies. Expansion of the irrigated area into new areas is not included in this model.

Table 5.1 is the summary of current efficiency used in the model for different provinces. In general, canals have higher efficiency than watercourse channels, since most canals are constructed using impervious concrete. Thus, the average system-wide efficiency is about 35 percent. Since canal and watercourse efficiency is a critical factor related to how much water can be diverted to the field, an efficiency improvement scenario will be discussed in a later section to evaluate the system-wide impacts of efficiency improvement.

Table 5.1 Typical Canal, Watercourse, and Field Efficiency for Different Provinces
percent

Province	Canal	Watercourse	Field
NWFP	0.76	0.59	0.89
Punjab	0.74	0.54	0.87
Sindh	0.80	0.52	0.85
Baluchistan	0.78	0.61	0.83
Indus system-wide	0.76	0.55	0.86

Note: NWFP = North-West Frontier Province.

Water input data include the surface water inflow, historical canal diversions, rainfall, evaporation and sub-irrigation and public tubewell pumping. Data (50-year record) from the nine inflow tributaries into the model (Indus, Chenab, Jhelum, Harro, Kabul, Ravi, Soan, Sutlej, and Swat) are used. The subsequent analysis uses different exceedance probabilities of inflow to assess the system-wide impacts and sensitivity to inflow. The baseline run uses the 50 percent exceedance probability, which equals 132 million acre-feet (MAF) annually available to the IBIS.

The total live storage in the system is about 11.5 MAF currently. Four reservoirs are used in the current model structure: Mangla, Tarbela, Chashma, and Chotiari. The Mangla dam is located on the Jhelum River. The current live capacity has declined to 4.6 MAF from the designed 5.3 MAF due to sedimentation. Tarbela dam on the Indus River is the largest earth-filled dam in the world. Completed in 1974, it was designed to store water from the Indus River for irrigation, flood control, and hydroelectric power generation. The designed storage is 11.3 MAF and the existing live storage is 7.3 MAF. The sedimentation rate is lower than expected, so the lifespan is re-estimated as 85 years to 2060. The Chashma barrage is located on the Indus River. Unlike Mangla and Tarbela, it is not primary used for irrigation but for power generation. The installed capacity of the power station is 184 megawatts (MW), comprising eight bulb-type turbine units of 23 MW capacity each. Chotiari Reservoir is an off-stream reservoir on the Nara and Lower Indus Canal systems. It was built to store water and irrigate the Thar Desert. Both Chashma and Chotiari have live capacity of less than 1 MAF.

Equations and Constraints
In total, 26 General Algebraic Modeling System (GAMS) equations are used in the IBMR to optimize the complex process related to water allocation and economics activities. These equations are categorized into six classes: (1) objective function, (2) economic equations, (3) water balance equations, (4) canal equations, (5) crop equations, and (6) livestock equations. This section will highlight some of the key equations and constraints. The detailed discussion of these equations can be found in Ahmad, Brooke, and Kutcher (1990).

Objective Function
The objective function of the IBMR is to maximize the CPS for the entire basin as described in figure 5.1 and given in equation (5.1). The objective function is only for the agriculture sector and does not include hydropower production or municipal and industrial water consumption. The primary decision variables are production across the agriculture commodity, groundwater type (saline or fresh), and ACZs.

$$CPS = \sum_Z \sum_G \sum_C Price_{Z,C} \times Production_{Z,G,C} - \sum_Z \sum_G Cost_{Z,G}$$
$$- \sum_Z \sum_C Import_{Z,C} - ImaginaryWater + \sum_Z \sum_C Export_{Z,C} \quad (5.1)$$
$$+ \sum_M \sum_N WaterValue_{M,N}$$

where Z is index for ACZ, G is index for groundwater type, C is index for crop, M is index for month, and N is index for node or reservoir. *Price × Production* is the total benefit from crop production and livestock production. *Cost* is the total cost for production, *Import* is the total cost for importing crops, *Export* is the total benefit for exporting crops, and *WaterValue* is the value of water that flows to the sea (and not utilized in the system) or stored in reservoirs. This value can be set to reflect the economic benefit of maintaining environmental flows to the sea.

The *ImaginaryWater* parameter in the objective function represents the penalty when there is insufficient water in the network flow model. In reality, production would not cease if there is a shortage of irrigation water. Thus, to prevent infeasible solutions (that is, hard constraints on the water demands) this variable acts as a penalty against the objective value. Therefore, the cost of production will be higher under these circumstances. When this variable is non-zero, it indicates that full irrigation demands are not being met.

Cost Function

The cost function (equation 5.2) contains all the cost for farm crop and livestock production in each ACZ as shown:

$$Cost_{Z,G} = \Sigma_Z \Sigma_C \Sigma_S \Sigma_W (FERT_{Z,C,S,M} + MISCCT_{Z,C,S,M} + SEEDP_{Z,C,S,M}$$
$$+ TW_{Z,C,S,M} + TRACTOR_{Z,C,S,M}) + \Sigma_Z \Sigma_G \Sigma_A Animal_{Z,G,A}$$
$$+ \Sigma_Z \Sigma_{SEA} PP_{Z,SEA} + \Sigma_Z \Sigma_G \Sigma_M Labor_{Z,G,M} \qquad (5.2)$$

where S is the index for cropping sequence (for example, standard, late, or early planting); W is the index for water stress (for example, standard, light, or heavy stress); A is the index for different animals (cow, bullock, and buffalo); and SEA is the index for season (rabi and kharf). *FERT* is the cost for fertilizer, *MISCCT* is the miscellaneous cost like insecticides and herbicides, *SEEDP* is the cost for seed, *TW* is the energy cost for groundwater pumping, *TRACTOR* is the cost for operating tractors, *Animal* is the fixed cost for livestock, *PP* is the cost for purchased protein concentrates for animals, and *Labor* is the cost for hiring labor.

Surface Water Balance Equation

Water balances in the river network and root zone are the essential mass balances in the IBMR. The surface water balance is related to the river routing process in the IBMR. Equation (5.3) describes the entire river network monthly water balance at each node:

$$\Sigma_I Inflow_I^M + \Sigma_N RIVERD_N \times TRIB_N^M + \Sigma_N RIVERC_N \times TRIB_N^{M-1}$$

$$+ \Sigma_N RIVERB_N \times F_N^M + \Sigma_N RIVERC_N \times F_N^{M-1} + \Sigma_N RCONT_N^{M-1}$$

$$- RCONT_N^M + Prec_N^M + Evap_N^M - \Sigma_N CANALDIV_N^M + ImaginayWater_N^M = 0$$

$$(5.3)$$

where I is the index for inflow node; *Inflow* is the streamflow; *RIVERD* is the routing coefficient for tributaries; *TRIB* is the tributaries' flow; *RIVERC* is the routing coefficient for previous month; *RIVERB* is the routing coefficient for mainstream; F is the mainstream's flow; *RCONT* is the monthly reservoir storage; *Prec* is the rainfall at reservoir; *EVAP* is the evaporation loss at reservoir; *CANALDIV* is the canal diversion; and *ImaginaryWater* is the imaginary surface water needed at nodes.

The root zone water balance at each ACZ in the IBMR is the relationship between the total available water in the root zone and the total crop water requirements, as shown in figure 5.5. The following equation (5.4) describes this balance.

$$Max[(WNR^M_{Z,G,C,S,W} - SUBIRRI^M_{Z,G} \times LAND^M_{Z,G,C,S,W}),0] \times X^M_{Z,G,C,S,W}$$

$$\leq TW^M_{Z,G} + GWT^M_{Z,G} + WDIVRZ^M_{Z,G} + ImaginaryRWater_{Z,G} \qquad (5.4)$$

where *WNR* is the water requirement from crops, *SUBIRRI* is the sub-irrigation, X is the cropped area, *TW* is the total private tubewell pumping, *GWT* is the public tubewell pumping, *WDIVRZ* is the surface water diversion and *ImaginaryRWater* is the imaginary water at the root zone.

Major Constraints

Canal capacity: The physical canal capacity (of the existing IBIS infrastructure) is used as the upper boundary of canal water diversion in the model. That is, expansion of the irrigated area (beyond the cultivable command) is not considered in this work. Note that this model was previously used to examine the cost-benefit of new irrigated areas as part of WAPDA's Water Sector Investment Planning Study (WAPDA 1990).

Provincial Historical Diversion Accord: Maintaining the 1991 Provincial Accord (described in chapter 2) is another constraint in the model. This water sharing agreement specifies how much water is to be diverted to each province (table 2.1). In order to consider this Accord in the IBMR, the actual monthly canal-wise diversion data from 1991 to 2000 are averaged and utilized as the constraint itself ("DIVACRD"). That is, it is assumed that the average historical diversions to each canal command (intra-provincial allocations) in the aggregate meets the inter-provincial requirements. In an earlier version of the model, this DIVACRD was set as an equality constraint. However, this eliminates the possibility for intraprovincial optimization. In this study, a 20 percent deviation from the monthly canal diversion was allowed, that is, each canal command diversion can be 0.8–1.2 percent of this historical value (while maintaining the physical constraints in the system and the aggregate provincial allocations).

Reservoir operation rule: No complex operation rules have been applied to these reservoirs. Only the upper and lower boundaries of reservoir storage have been set up. This is acceptable given that the model operates for a single year.

Output Data

Every item in the objective function is an output: revenue from production, farm cost, imports, exports, total welfare, value of water in the reservoir, and the flow to the sea. The imaginary parameters can also be checked to see if their values are non-zero (thus, indicating a shortfall in demanded irrigation waters).

The major output that will be examined in this study is the cropped areas for different crops, ACZs, and months. The model also provides detailed information for every combination of cropped area. For example, production can be summed across ACZs or provinces or from monthly to seasonal and annual time scales. The results are also given in each ACZ for areas that are sourced from different groundwater types (fresh and saline). Resources used, such as labor and private tubewells, can also be calculated for each ACZ. The power generation from reservoirs is a by-product from the model and calculated after an optimized solution is found. Finally, the surface water and groundwater balance is determined from the calculations of the monthly canal diversion; monthly node-to-node flow balance; reservoir inflow, outflow, and storage; surface water to the root zone for each ACZ; and groundwater depth and recharge.

Baseline: Year 2008/09

This section presents the baseline performance of the IBMR-2012. All scenario simulations will be in relation to this baseline. Table 5.2 and figure 5.6 show the major outputs from the baseline model. The basin system-wide objective value is PRs 2,850 billion (US\$35.62 billion[1]). This is consistent with the agriculture gross domestic product (GDP) for the entire country of US\$34.8 billion (World Development Indicators). Punjab has the largest cropped area, followed by Sindh. Surface and groundwater use across the provinces follow closely the DIVACRD constraint. Most groundwater usage is in Punjab. According to NTDC (2010), the annual generated power in 2009 at Tarbela, Mangla, and Chashma was 13.95, 4.79, and 1.09 billion kilowatt-hours (BKWH), respectively. The total power generation is 19.83 BKWH and is consistent with the modeled results.

Table 5.3 shows the revenues across different farm commodities, which vary by province. Basmati, cotton, sugarcane, and wheat generate the most revenue

Table 5.2 Major IBMR Outcome under Baseline Condition

	Objective value (PRs, millions)	Commodity total revenue (PRs, millions)	On-farm costs (PRs, millions)	Cropped area (1,000 acres)	Crop production (1,000 tons)	Power generation (BKWH)
Indus	2,850,099	3,162,371	601,369	48,491	95,138	19.59
Punjab	n.a.	2,430,117	440,965	34,734	65,374	n.a.
Sindh	n.a.	628,036	132,823	11,057	24,905	n.a.
Other		104,218	27,582	2,701	4,859	n.a.

Note: BKWH = billion kilowatt-hours, n.a. = not applicable. "Other" includes Balochistan and North-West Frontier Province (NWFP).

Figure 5.6 Model Irrigation Use by Source and Province
million acre-feet

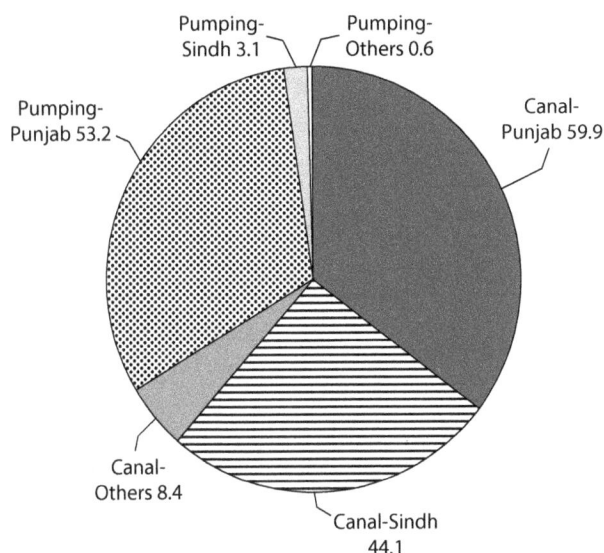

Table 5.3 Commodity Revenue Breakdown for the Baseline Condition
PRs, millions

	Pakistan	Punjab	Sindh	Others
Basmati	749,694	749,694	0	0
Irrigated rice	170,466	27,733	108,530	34,204
Cotton	674,609	552,092	122,190	327
Gram	36,101	20,860	12,810	2,431
Maize	70,457	44,542	692	25,223
Mus+rap	2,574	1,923	7	645
Sc-mill	245,950	156,249	78,764	10,937
Wheat	418,049	377,301	35,080	5,669
Potatoes	111,421	108,316	682	2,424
Onions	56,891	18,360	37,187	1,344
Chili	35,674	17,685	17,962	27
Cow-milk	144,051	76,608	64,160	3,282
Buff-milk	446,434	278,755	149,973	17,706
Total	3,162,371	2,430,117	628,036	104,218

in Punjab. Irrigated rice and cotton revenues are highest in Sindh. These baseline numbers are consistent with actual agriculture census data for 2008 (see appendix B). Table 5.4 breaks down the on-farm cost for different categories. In general, the primary production costs are labor hired, tractor, and fertilizer use. The following climate change impact and investment analysis will be compared against these results.

Table 5.4 Farm Cost Breakdown under Baseline Condition
PRs, millions

	Pakistan	Punjab	Sindh	Others
Seed	38,434	27,035	9,189	2,210
Labor	205,834	145,040	51,970	8,824
Water	47,343	37,711	8,756	876
Protein	2,488	1,988	401	98
Fertilizer	120,082	79,972	33,261	6,849
P-well	45,930	42,942	2,530	458
Livestock	735	432	273	30
Tractor	140,524	105,844	26,442	8,238

Figure 5.7 CGE Conceptual Diagram

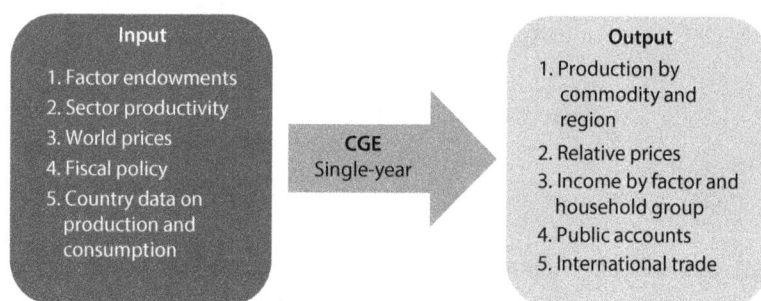

Input

1. Factor endowments
2. Sector productivity
3. World prices
4. Fiscal policy
5. Country data on production and consumption

CGE
Single-year

Output

1. Production by commodity and region
2. Relative prices
3. Income by factor and household group
4. Public accounts
5. International trade

Note: CGE = computable general equilibrium.

Section 2: Computable General Equilibrium Model

This section provides a brief description of the CGE model. Detailed information can be found at Diao et al. (2011). The inputs for the CGE model include factor endowments (amount of labor, land, and capital); sector-specific productivity; world prices; fiscal policy (tax rates, government expenditure); and updated country-specific data on production (value-added and intermediate use by sector) and consumption (value of consumption for each commodity and household group). The outputs of the CGE include production by commodity and region; relative prices; income by factor and household group; public accounts (for example, public deficit); and international trade (exports and imports by commodity). Figure 5.7 is a conceptual diagram that explains the modeling process of the CGE.

CGE Model Structure

In essence, the CGE model takes into account the interaction between producers and consumers in the economy. The model tracks the selling of goods from firms to households and to other firms, the selling of factor services from households to firms, and the generation of savings that finance the investment in the economy, as shown in figure 5.8. The arrows in the figure track the (explicit or implicit) payments in the countrywide economy. Firms pay wages and rents to

Figure 5.8 Flow in the CGE Model

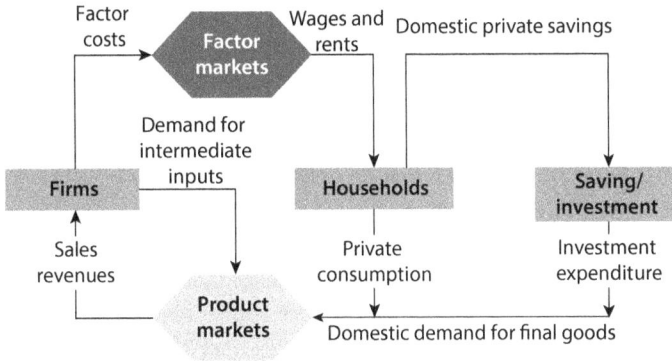

Note: CGE = computable general equilibrium.

households, and they buy goods produced by other firms (intermediate inputs). Households use their income to buy products from firms and to save, in turn financing investment, that is, an additional demand for firms' product markets. The components of the model follow, going through the behavior of consumers and producers; the introduction of the government, investment, and the foreign sectors; the way the model equilibrates supply and demand in the product and factor markets; and its macro-economic behavior and regional disaggregation.

Consumer and Producer Behavior

Following general equilibrium theory, representative consumers (that is, households) and producers in the model are treated as individual economic agents. Household incomes are determined by the sum of factor income, public transfers, and foreign transfers. Households use their income for consumption and saving. Consumption is allocated over different goods to maximize household's utility[2] subject to their budget constraints. Producers are defined at the sector level (that is, agriculture, industry, services). Each representative producer maximizes profits, taking the prices at which they buy/hire inputs and sells their outputs as given. Following neoclassical theory, constant returns are assumed to scale in production, that is, increasing all the factors by a given percentage leads to increasing the production by the same percentage. In particular, a constant elasticity of substitution (CES) function is used to determine production:

$$X_i = \Lambda_i (\Sigma_f a_{if} \cdot V_{if}^{-\rho_i})^{-1/\rho_i} \tag{5.5}$$

where X is the output quantity of sector i, Λ is a shift parameter reflecting total factor productivity (TFP), V is the quantity used of each factor f (that is, land, labor, and capital) by the firm, and α is a share parameter of factor f employed in the production of good i. The maximization of profits by the sector-specific producers provides the system of factor demand equations used in the model. For example, if the amount sold from a producer increases, it will make him try to hire more workers and acquire more necessary factors for his production.

The model captures the use of intermediate inputs in the production process. In particular, the demand for intermediates is based on fixed physical input-output coefficients.

Government and Investment Demand

The government is treated as a separate agent, with income and expenditures but without any behavioral functions. Total domestic revenues R is the summation of all individual taxes (sales taxes, income taxes, tariffs, and so on). Tax rates are exogenous so that they can be used to simulate policy changes. The government uses its revenues to purchase goods and services (that is, recurrent consumption spending), to make transfers to other actors (households, firms, non-residents) and to save (that is, finance public capital investment). The participation of the government demand in the domestic absorption of the economy is fixed in the model.

There are also no behavioral functions determining the level of investment demand for goods and services. The total value of all investment spending must equal the total amount of investible funds I in the economy. This value is split among different commodities in fixed proportion, as informed by the composition of investment in the National Accounts.

International Trade

Given observed two-way trade between countries for similar goods, we assume imperfect substitution between domestic goods and goods supplied to and from foreign markets. World prices are exogenous, reflecting the perception that the domestic economy does not have significant power to affect the world prices (small country assumption).

Equilibrium Conditions

Full employment and factor mobility across sectors is assumed for labor, and fixed sector-specific employment is assumed for land and capital. This means that when the price of a commodity increases, the producer of the commodity will increase its demand for factors, but only workers (who are mobile) will be able to move into the sector, while the other factors (which are fixed in the sectors they are in) will increase their wage in this sector. Relative prices are determined such that their supplies equal their demands. More specifically, in each commodity, the supply of the good Q equals total demand, composed of consumption by households (C_{ih}), investment (N_i), public sector demand (G_i), intermediate demand by other production sectors ($\Sigma_{i'}(io_{i'i} \cdot X_i)$, and net exports ($E - M$):

$$Q_i = \Sigma_h C_{ih} + N_i + G_i + \Sigma_{i'}(io_{i'i} \cdot X_i) + E - M \tag{5.6}$$

Macro-Economic Closures

Macroeconomic balance in a CGE model is determined exogenously by a series of "closure rules." The most important of these is the current account balance. While this is a substantive research topic within macroeconomics, it is treated as an exogenous variable within the single-country open economy CGE model. Either total savings S or total investment I, but not both, should be determined exogenously. This choice is called the "savings-investment" closure. The model is savings driven, with households saving a fixed share of their income and investment I automatically determined by the level of total available savings. Finally, recurrent consumption spending of the government G is a fixed share of domestic absorption, and public savings are endogenously determined by the model. Finally, the original consumer basket is chosen as the model's numeraire that is, the consumer price index (CPI) is fixed.

Regional Production

We disaggregate representative producers and households across sub-national regions, an extension that allows us to reflect spatial heterogeneity in geographic conditions. These are important considerations for agriculture, which depends on agro-ecological and climatic conditions, and for developing countries, where markets are often underdeveloped. Each regional producer has his own production function and associated technology coefficients, and often uses region-specific factors, such as agricultural land. Each regional producer supplies his output to a national product market, with output from each region combined into a composite national good through a CES aggregation function. Households are classified into groups that consider the region where they reside.

CGE Model Social Accounting Matrices Update

One of the main advantages of CGE models over theoretical models is their calibration to detailed empirical data. "Calibration" refers to the process of assigning values to the model's parameters and variables, typically using observed country data. Some of the assumptions that the authors made when specifying the CGE model were done to ease its calibration, since in many cases the data needed for more complex functional forms is unavailable in developing countries. For example, the reason a function is used that is based on constant income elasticities to determine consumer demand is because it requires data that can readily be obtained from household surveys (that is, expenditure shares and income elasticities). More elaborate functions often drop this constant income elasticities assumption, such as in the "Almost Ideal Demand System," but need more detailed data. Calibrating the behavior of more complicated functional forms often involves simply making more assumptions where data are unavailable. The next section describes the data sources and estimation procedures used to calibrate the CGE model.

Social Accounting Matrices

The values of almost all variables and parameters in the CGE model are drawn from a social accounting matrix (SAM; Pyatt and Round 1985; Reinert and Roland-Holst 1997). Constructing a SAM is therefore a fundamental part of developing a CGE model for a country. A SAM is an economy-wide representation of a country's economic structure. It captures all income and expenditure flows between producers, consumers, the government, and the rest of the world (ROW) during a particular year. Table 5.5 presents the basic structure of a SAM that could be used to calibrate the core model described above. The SAM contains a number of "accounts" representing different agents in the model, including sectors (producers) and households (consumers). The rows and columns of the SAM represent incomes and payments, respectively, from one account to another. As with double-entry accounting, the SAM is a consistent economy-wide database because row and column totals must be equal. In other words, a payment from one account always becomes an income for another. The SAM therefore provides the baseline year equilibrium state for the CGE model.

A SAM is constructed in two stages. During the first stage, data from different sources are entered into each of the SAM's cells. As with the CGE model, the SAM allows for multiple sectors and households. Thus, the "sector", "product," and "household" rows and columns actually contain many subaccounts. The three main data sources for constructing a SAM are national accounts, input-output tables (or supply-use tables), and nationally representative household budget surveys. As shown in table 5.5, national accounts provide information on the composition of GDP at factor cost (that is, sectoral value-added) and by broad expenditure groups at market prices. The technical coefficients (that is, the requirements of inputs produced by other industries to produce a given commodity) in the input-output table are scaled in light of the value-added in the given commodity to estimate intermediate demands. (That is, the matrix that captures the use of goods by a sector of goods produced by other sectors and the same sector).

The SAM also disaggregates government and investment demand across products. The household survey is used to segment labor markets (that is, disaggregate labor income into different groups, such as by education). The survey also defines households' expenditure patterns and the distribution of incomes to representative household groups. Therefore, the survey data is the main determinant of differential income and distributional effects across household groups in the CGE model.

Other databases are used to complete specific cells within the SAM. Government budgets provide information on tax rates, revenues, and expenditures. Although not shown in table 5.5, government budgets (and household surveys) also determine the level and distribution of social transfers (that is, payments from government to households, like pensions and subsidies to the poor). Customs and revenue authorities provide data on imports and exports and their associated tariffs and subsidies. The balance of payments, usually compiled by a country's central bank, is used to populate the external or ROW account, including information on transfer receipts and payments and the current account balance. Finally, sectors in the SAMs are usually disaggregated across subnational regions using information on regional production and technologies from agricultural and industrial surveys. The information about regional production is a key link between the CGE and IBMR models: the CGE takes as input the ratio between production and area that come out of IBMR to update the (total factor) productivity in the CGE model, which is an exogenous parameter. Trade margins,

Table 5.5 General Structure of a Social Accounting Matrix

	Sectors	Products	Factors	Households	Government	Investment	Rest of the world	Total
Sectors		Marketed supply (PD, D)					Export demand (PE, E)[f]	
Products	Intermediate demand (io)[b]			Private consumption (C)[a, c]	Public consumption (G)[a, b, d]	Investment demand (N, ε)[a, b]		Total demand
Factors	Value-added (V, W, Z)[a, c]							Factor income
Households			Income distribution (δ)[c]				Transfers (hw)[c, e, f]	Household income (Y)
Government	Indirect tax (te)[d, f]	Indirect tax (tc, tm)[d, f]	Factor tax (tf)[d]	Income tax (ty)[c, d]			Transfers (rw)[d, e, f]	Total revenues (R)
Savings				Private savings (s)[a, c]	Public savings (FB)[a, d]		Foreign savings (FS)[e]	Total savings (S)
Rest of world		Import supply (PM, M)[f]						Total foreign payments
Total	Gross output (PP, X)	Total supply (P, Q)	Factor payments	Total household spending	Recurrent spending	Total investment (I)	Total foreign receipts	

Note: Main data sources used to populate the SAM: a. national accounts and regional production data; b. input-output tables and industrial surveys; c. household and labor force surveys; d. government budgets; e. balance of payments; and f. customs data and tax revenue authorities.

which are not shown in the table, are estimated using information on producer and consumer prices. Trade margins may also be drawn from input-output or supply-use tables.

Inevitably, inconsistencies occur between data from different sources, which lead to unequal row and column totals in the model's SAMs. Therefore, the second stage of constructing a SAM is to "balance" these totals. This reconciliation of data from disparate sources is similar to a "rebasing" of national accounts. Cross-entropy econometric techniques are used to estimate a balanced SAM (Robinson, Cattaneo, and El-Said 2001). This is a Bayesian approach that uses a cross-entropy distance measure to minimize the deviation in the balanced SAM from the unbalanced prior SAM containing the original data. Constraints such as total and sectoral GDP are imposed during the estimation procedure to reflect narrower confidence intervals around better-known control totals (for example, total GDP). The SAM and its underlying data sources provide almost all of the information needed to calibrate the CGE model. Only the behavioral elasticities remain.

Behavioral Elasticities and Other External Data

Behavioral elasticities are needed for the consumption, production, and trade functions. The demand function requires information on income elasticities and the marginal utility of income with respect to income (Frisch parameter, see Frisch 1959). Marginal budget shares (the fraction of consumption that a given household allocates to each commodity) are derived by combining the estimated income elasticities with the average budget shares drawn directly from the SAM. The income elasticities in this case are based on a set of priors given by the price elasticities of different crops from IBMR and a cross-entropy process. Trade elasticities determine how responsive producers and consumers are to changes in relative prices when deciding to supply goods to or purchase goods from foreign markets. Higher elasticities are expected when substituting between more homogenous products, such as maize and copper. Lower elasticities are expected for more differentiated product categories, such as chemicals and machinery. In Pakistan, most of the Armington elasticities (the elasticity of substitution between domestic and imported commodities) are in the range of 2–3, with the following exceptions: chemicals (0.5), cement (0.5), petroleum (0.8), and manufacturing (0.5). In most developing countries, the data needed to econometrically estimate country-specific elasticities do not exist—at least not in an appropriate form (Arndt, Robinson, and Tarp 2002). The elasticities governing factor substitution in the production rarely exist for developing countries. In the absence of reliable country-specific estimates, we assume inelastic factor substitution for most activities in the [0.75, 0.90] range.

Social Accounting Matrix for Pakistan

Building an updated SAM for Pakistan was started from Pakistan SAM 2001–02 developed by Dorosh, Niazi, and Nazli (2006). Given that the most recent available input-output table is for 1990–91, and the published national accounts on an even earlier input-output table (1985), in preparing the 2001–02 Pakistan SAM it was necessary to construct a consistent set of accounts for production and value-added by sector based on the 1991 input-output table. These accounts then formed the base upon which factor and household accounts were disaggregated.

Four major types of accounts are distinguished in the 2001–02 Pakistan SAM: (1) activities, (2) commodities, (3) factors of production, and (4) institutions (households, government, and the rest of world), including an aggregate institutional savings-investment account, which collects all the savings in the economy and uses them to finance investment. The production accounts describe the values of commodities (goods and services) and inputs into each production activity along with payments to factors of production (land, labor, water, and capital). Commodity accounts show the components of total supply in value terms, domestic production, imports, indirect taxes and marketing margins; and total demand, intermediate input use, final consumption, investment demand, government consumption, and exports. Factor accounts describe the sources of factor income (value-added in each production activity) and how these factor payments are further distributed to the various institutions in the economy (households of different types, enterprises, government, and the ROW). Accounts for institutions record all income and expenditures of institutions, including transfers between institutions. Savings of the different institutions and investment expenditures by commodities are given in the savings-investment accounts.

Data Used to Build the 2008 Pakistan SAM

The 2008 SAM uses data mainly from the following sources:

- 2008 Macroeconomic Aggregates
- 2001–02 National Accounts (value-added for 15 sectors)
- 1990–91 Input-Output Table (97 sectors)
- 2001–02 Pakistan Integrated Household Survey (consumption disaggregation)
- 2001 Pakistan Rural Household Survey (household income disaggregation)
- 2001–02 Pakistan Economic Survey (sector/commodity data on production, prices, trade)

Structure of the 2008 Pakistan SAM

The 2001–02 Pakistan SAM will now be updated into a 2008 Pakistan SAM. (Appendix C lists the accounts of the 2008 Pakistan SAM). The SAM includes 63 activities, 36 of which are national and the remaining, regional. For the agricultural activities, returns to land and own-family labor are disaggregated by region (Punjab, Sindh, and Other Pakistan) and by size of farm: small (0–12.5 acres), medium (12.5–50 acres), and large (50 acres plus) farms (defined according to area cultivated, not land ownership). Of the 27 factors of production specified, 23 involve only agricultural production: 8 types of agricultural labor; 12 types of land, water, livestock capital, and 3 types of other agricultural capital.

This detailed treatment of rural factors and agriculture in the SAM reflects the primary objective of constructing the SAM: to better understand the relationship between agricultural performance and rural income growth in the context of imperfect rural factor markets. Fifteen of the 19 household categories are rural agricultural households, split according to amount of land cultivated (large farm, small farm, landless) and region (Sindh, Punjab, and Other Pakistan). Non-farm households, both rural and urban, are split into poor and non-poor, according to their 2000–01 per capita household expenditures, with poor households defined as those with a per capita expenditure of less than 748 PRs per month per capita (22.67 percent of urban households). Non-farm rural households, defined as rural households for which the main occupation of the head of household is not crop or livestock farming, form the last household group, accounting for 19.8 percent of total population. The structure and accounts in the 2008 Pakistan SAM can be found in appendix C.

Integration of the Two Models

As this brief description shows, the CGE-SAM modeling approach can provide a much wider range of agro-economic linkages than can be addressed with the IBMR. The IBMR and CGE as described here have not been jointly used to date, and there are interesting analytical challenges in linking them (refer to Yu et al. 2010, for a comparable study in Bangladesh). There are many potential linkages between the IBMR and CGE models. In this study, IBMR outputs are exported to the CGE model to compute a wider and more complex array of potential agro-economic impacts in Pakistan. Two of the primary outputs from the IBMR are crop production and cropped area. To evaluate the impact of climate change and the effect of adaptation investments on GDP, Ag-GDP and other economic outputs, crop production and cropped area are passed directly to the CGE model. The production and cropped area for the baseline and subsequent climate and investment scenarios are used to update the (total factor) productivity in the CGE model. All results from the CGE are relative to the 2008 baseline. The results from these analyses are presented in chapter 6. Figure 5.9 illustrates how the two models are integrated.

Model Limitations

The models used here are among the best mathematical representations available of the physical and economic responses to exogenous future climate risks. Both IBMR-2012 and CGE-2008 are single-year models. That is, a comparative statics approach is taken here in examining future scenarios instead of a dynamic approach. In IBMR-2012, a simple measure for environmental consideration is used; groundwater dynamics are limited; energy production is a by-product of the system optimization; only one dimension of food security, food self-supply, is examined; no flooding damage is considered and no detailed cost-benefit analysis of investment scenarios are conducted. In CGE-2008, only one composition

Figure 5.9 Integration of IBMR and CGE

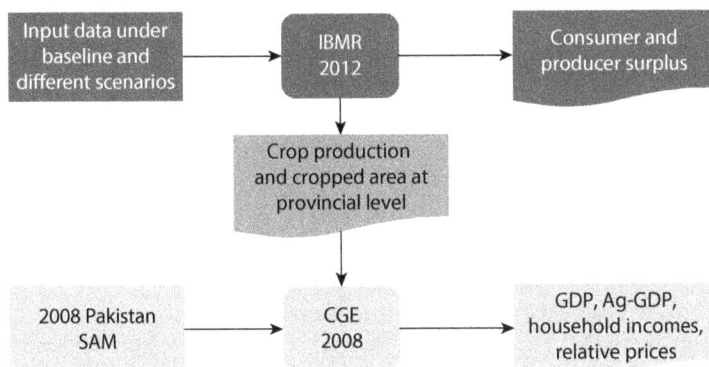

Note: CGE = computable general equilibrium, IBMR = Indus Basin Model Revised, SAM = social accounting matrix.

of Pakistan economy (current condition) is used; the world sugar, wheat, and other prices would not change due to climate change in the future. As in all modeling approaches, uncertainty exists as to parameters that may not be known with precision and functional forms that may not be fully accurate. Thus, careful sensitivity analysis and an understanding and appreciation of the limitations of these models are required. Further collection and analysis of critical input and output observations (for example, snow and ice data) will enhance this integrated framework methodology and future climate impact assessments.

Notes

1. US$1 = PRs 80 (2009).
2. In particular, through a Stone-Geary utility function.

References

Ahmad, M., A. Brooke, and G. P. Kutcher. 1990. *Guide to the Indus Basin Model Revised.* Washington, DC: World Bank.

Ahmad, M., and G. P. Kutcher. 1992. "Irrigation Planning with Environmental Considerations: A Case Study of Pakistan's Indus Basin." World Bank Technical Paper 166, World Bank, Washington, DC.

Arndt, C., S. Robinson, and F. Tarp. 2002. "Parameter Estimation for a Computable General Equilibrium Model: A Maximum Entropy Approach." *Economic Modeling* 19 (3): 375–98.

Diao, X., P. Hazell, D. Resnick, and J. Thurlow. 2011. *Agricultural Strategies in Africa: Evidence from Economywide Simulation Models.* Washington, DC: IFPRI.

Dorosh, P., M. K. Niazi, and H. Nazli. 2006. "A Social Accounting Matrix for Pakistan, 2001–02: Methodology and Results." Working Paper, Pakistan Institute of Development Economics (PIDE), Islamabad.

Frisch, R. 1959. "A Complete Scheme for Computing All Direct and Cross Demand Elasticities in a Model with Many Sectors." *Econometrica* 27 (2): 177–96.

NTDC (National Transmission and Dispatch Company). 2010. *Electricity Marketing Data—Power Systems Statistics, 35th issue*. WAPDA House, Lahore, Pakistan.

Pyatt, G., and J. I. Round, eds. 1985. *Social Accounting Matrices: A Basis for Planning*. Washington, DC: World Bank.

Reinert, K. A., and D. W. Roland-Holst. 1997. "Social Accounting Matrices." In *Applied Methods for Trade Policy Analysis*, edited by Joseph F. Francois, 94–121. Cambridge, U.K.: Cambridge University Press.

Robinson, S., A. Cattaneo, and M. El-Said. 2001. "Updating and Estimating a Social Accounting Matrix Using Cross Entropy Methods. *Economic Systems Research* 13 (1): 47–64.

WAPDA (Water and Power Development Authority). 1990. *Water Sector Investment Planning Study (WSIPS)*. 5 vols. Lahore, Pakistan: Government of Pakistan Water and Power Development Authority.

Yu, W., M. Alam, A. Hassan, A. S. Khan, A. C. Ruane, C. Rosenzweig, D. C. Major, and J. Thurlow. 2010. *Climate Change Risks and Food Security in Bangladesh*. Earthscan, London, U.K.

CHAPTER 6

Sensitivity and Scenario Results

Key Messages

- The integrated model is most sensitive to inflows into the system, crop water requirements, and the depth to groundwater.
- The water allocations per the 1991 Provincial Accord and within provinces are the most critical constraint in the Indus system. By relaxing the Accord constraint and allowing optimal economic allocation between and within provinces, both Punjab and Sindh provinces stand to gain. The ability to manage extreme events (for example, drought) by more reliably meeting system-wide demands is also enhanced.
- Climate futures were examined representing a plausible range of climate changes within the next 80 years consistent with recent observations and theory.
- Gross domestic product (GDP), Ag-GDP, and household income are estimated to decrease by 1.1, 5.1, and 2.0 percent, respectively, on an annual basis as a result of plausible climate changes. In the most extreme future—when inflow is 90 percent exceedance probability and the temperature increases +4.5°C—GDP, Ag-GDP, and household income are estimated to decrease by 2.7, 12.0, and 5.5 percent, respectively, on an annual basis.
- Climate impacts on crop production are greatest in Sindh (–10 percent on average).
- Irrigated rice, sugarcane, cotton, and wheat demonstrated the greatest sensitivity to climate, and changes represent both response to climate and dynamic responses to water availability and price changes. Milk revenues are also expected to decrease.
- Three possible adaptation investments were evaluated: improvements to system-wide efficiency, construction of new storage, and investments in agriculture technologies to increase crop yield.
- From a system perspective, additional storage provides agricultural benefits by mitigating the effects of droughts, but it provides little additional agricultural benefit (assuming no expansion of the current irrigated area) in other years.

This is at least partially due to the current constraints on agricultural production, including allocation constraints such as the 1991 Accord.

- Although the model does not optimize for hydroelectricity production, additional storage does result in increased hydropower and consequent economic benefit. Flood risk reduction was not considered in this report but is potentially significant.
- Canal efficiency and crop yield investments show potential to minimize the impacts of future climate risks and meet food self-sufficiency objectives, increasing production by 5–11 percent on average and offsetting future climate losses.
- Without specific interventions, environmental considerations, such as flow to the sea, changes in depth to groundwater, and the overall salinity situation, are projected to worsen. Potential adjustments to climate and food risks need additional investigation.

Sensitivities of Hydrologic Parameters and the DIVACRD Constraint

The most sensitive parameters in the Indus Basin Model Revised (IBMR) (figure 6.1a) are stream inflow into the Indus, crop water requirement, and depth to groundwater. The objective value can change significantly with different available inflows. The lowest total inflow tested (90 percent exceedance probability) is 101 million acre-feet (MAF) and the highest value (10 percent exceedance probability) is 209 MAF. The objective value decreases to almost 60 percent of the baseline (see chapter 5, section "Baseline: Year 2008–09") when the inflows drop to 101 MAF. When the inflow increases to 209 MAF, the objective value change is small (+0.1 percent from the baseline). That is, for high-flow settings, the system is unable to generate more economic benefits in the basin given the current constraints, including water allocation requirements from the 1991 Accord and physical capacity and land area constraints (that is, irrigated area served by the Indus Basin Irrigation System [IBIS] is fixed).

Increasing temperatures are expected to increase evaporative demand from crops and soils, which would increase the amount of water required to achieve a given level of plant production (Brown and Hansen 2008). The crop water requirement parameters in IBMR are based on theoretical consumptive requirements, survey data, and model experiments of water balances of the entire basin (Ahmad, Brooke, and Kutcher 1990). A local study by Naheed and Rasul (2010) is used to link crop water requirement and air temperature change under the assumption that crop phenology and management will remain the same under different air temperature conditions. The modeling results indicate that when the crop water requirement increases more than 5 percent above the baseline irrigation requirements (corresponding to a temperature increase larger than 2°C), the objective value drops significantly. Figure 6.1a shows that this temperature increase will result in a 42 percent decrease in the objective value (from the baseline). The highest tested crop water requirement is +35 percent more than the baseline which corresponds to a 6.5°C temperature increase.

Figure 6.1 IBMR Sensitivity Analysis Results

a. Sensitivity analysis results on hydrologic parameters

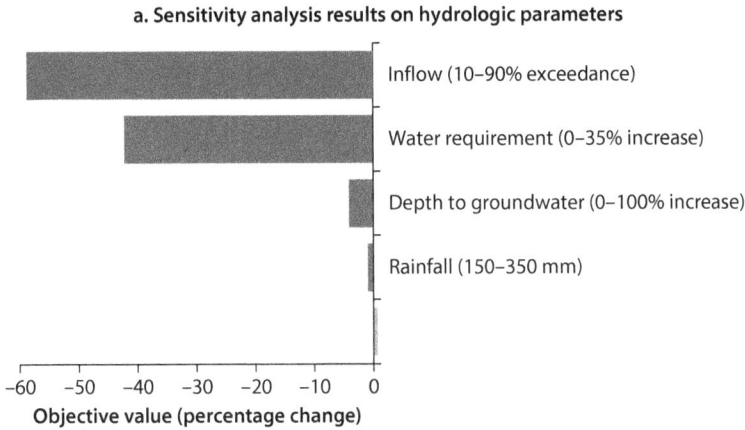

Inflow (10–90% exceedance)

Water requirement (0–35% increase)

Depth to groundwater (0–100% increase)

Rainfall (150–350 mm)

−60 −50 −40 −30 −20 −10 0

Objective value (percentage change)

Note: IBMR = Indus Basin Model Revised. Objective value baseline = PRs 2,850 billion.

b. Sensitivity analysis results on 1991 Provincial Water Allowance Accord (DIVACRD) constraint

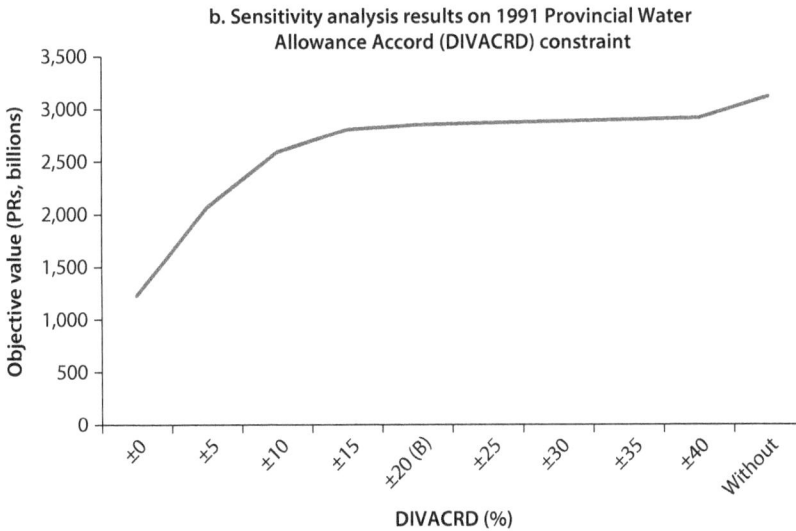

DIVACRD (%)

Note: Empty data points indicate the use of imaginary water. The "±" means tolerance range, the "(B)" means baseline and "Without" means runs without the Accord.

This is an unlikely temperature change in the next several decades but illustrative of the dynamics and sensitivity of the system.

The objective value is also sensitive to the depth to groundwater, which varies across agro-climatic zones (ACZs). Figure 6.1a shows that the objective value decreases by about 4 percent when average groundwater depth doubles throughout the system. Note that the unit pumping cost is constant (a function of volume only) and does not increase with depth. This is a limitation in the current model. Groundwater issues are discussed further in the "Environment Issues" section.

The historical canal diversion constraint (DIVACRD) simulates the 1991 Provincial Accord requirement (described in chapter 2). This water allocation

Table 6.1 Analysis of Impact across Provinces with and without the Provincial Accord
(PRs, millions)

Province	Revenue (PRs, millions)	Cost (PRs, millions)	Net revenue[a]	Profit change if remove DIVACRD	Canal diversion (MAF)	Change in canal diversion	Marginal (PRs/AF)
Fixed provincial allocation							
Punjab	2,390,054	433,072	1,956,982		55.8	n.a	n.a
Sindh	573,822	116,339	457,482		45.6	n.a	n.a
Others	104,218	27,257	76,961		9.3	n.a	n.a
Optimized allocation							
Punjab	2,503,663	463,216	2,040,447	83,464 (4%)	61.4	5.6	14,904
Sindh	718,849	178,782	540,067	82,584 (18%)	57.0	11.4	7,244
Others	103,913	34,038	69,876	−7,086 (−9%)	6.6	−2.7	−2,624

Note: n.a. = not applicable, DIVACRD = 1991 Provincial Water Allocation Accord, MAF = million acre-feet, AF = acre-foot.
a. Net revenue = revenue−cost.

constraint is the most critical constraint in the model. Figure 6.1b shows the objective value for varying levels of the constraint—from strict adherence to no constraint. For a ±x percent deviation, canal diversions can vary between a (1−x) to (1+x) fraction of the historical canal allocations. As the constraint is increasingly relaxed, more objective value (that is, economic benefit) is possible. The objective value ranges by a factor of 2. When the deviation allowed is smaller than ±15 percent, the objective value shows the largest changes. By fully relaxing this constraint, the largest objective value is achieved. Under these circumstances, the only binding constraint becomes the actual physical capacity of the system, both canal and land capacities. Note that for subsequent scenarios, a ±20 percent deviation is used as the baseline. This is the point at which sufficient irrigation water is available.

Table 6.1 shows the impact of DIVACRD across the different provinces. By completely relaxing this constraint, the optimized allocation results in additional canal water to both Punjab and Sindh and a consequent increase in economic benefits to both. In absolute terms, the additional net revenues in Punjab are estimated to be PRs 83,464 million (US$1.04 billion); for Sindh they are estimated to be PRs 82,584 million (US$1.03 billion in 2009). The increase in Sindh (18 percent) is larger in percentage terms than Punjab (4 percent). Moreover, a marginal analysis was done on this constraint over the entire basin which revealed that the incremental value-added per acre-foot (AF) is greatest in Punjab. This reflects the higher net returns from the agriculture sector in Punjab.

The aggregate gains from relaxing DIVACRD involve relative gains by some ACZs, canal commands, and crops as compared with others. This results from the allocation of water to its most economically productive uses at the ACZ level. Thus, the model simulates optimal economic allocation *both between and within provinces*. For example, while Punjab would gain a 5 percent increase in canal diversions with the relaxation of DIVACRD, the model also shows that within Punjab some ACZs would lose up to 5 percent. Thus, in order to implement and realize the full benefits of relaxing the 1991 Accord, consideration of how to provide incentives for winners and losers within provinces may be as important (or more) than the needed incentives between provinces, which for Sindh and Punjab are self-evident at the provincial level.

The potential benefits of relaxing the DIVACRD constraint look promising. However, these optimization results must be weighed in relation to the current state of interprovincial water relations and administration. As noted in chapter 2, the 1991 Accord was established to clarify interprovincial shares, that is, with the aim of increasing the reliability of provincial shares and deliveries, and thereby increasing the prospect for consensus on future infrastructure development of the sort envisioned by the Water Sector Investment Planning Study (WSIPS, WAPDA 1990). Unfortunately, neither aim has been sufficiently achieved. Briefly, Sindh did not trust Punjab's diversions before independence in 1947, let alone before the Accord of 1991 (see Michel 1967); neither province has trusted the other or Indus River System Authority (IRSA) under the Accord as currently administered; and presumably they would not expect to the other to take or receive their "optimal shares" under a relaxed Accord. IRSA's technical and administrative limitations have been discussed in previous studies (see review by Tariq and Ul Mulk 2005, for Briscoe and Qamar 2006).

Furthermore, it is worthwhile noting that the measurement of actual canal flows, watercourse diversions, and water uses remains uncertain and disputed. Provincial departments have sought in various ways to raise the empirical standards for monitoring irrigation water diversions and use. At the interprovincial level, however, Pakistan made an unsuccessful effort to install telemetry equipment to improve real-time data quality for deliveries under the 1991 Accord, which IRSA eventually had to abandon. There have been recent calls for renewed investment in an advanced, high-quality measurement system.

Finally, even though it is unlikely and probably unwise that the DIVACRD constraint should—by itself—be relaxed, there is room for flexible policy adjustments and mechanisms within the wider framework of the present Accord (for example, interprovincial exchange of surplus allocations, water banking, and leasing arrangements), which the IBMR modeling results suggest should be pursued on agro-economic grounds. These could include mechanisms within provinces for exchanging water for compensation and also mechanisms for exchange between provinces. The results suggest that there may be significant gains not only in terms of relaxing the provincial Accord but also in implementing economic allocation within provinces. In fact, neither is mutually exclusive, and the greatest gains would result from economic allocation at both levels.

Future Climate Risk Scenarios

Climate scenarios were developed to examine the effects of possible hydrological or climatic changes that may occur in the future. Given the low confidence in general circulation model (GCM) projections in this region (Immerzeel et al. 2011), a more robust approach would be to evaluate responses across a wide range of plausible climate futures. Note that some of the future scenarios include greater precipitation, but all feature warmer temperatures. A future year is not specified since the IBMR is a single-year model. All results are compared with a baseline based on current climate. Thus, results for investments are presented as percent changes to those same investments under the current climate. The baseline includes the DIVACRD constraint. Results are presented as box-whisker plots and show 1st, 25th, 50th, 75th, and 99th percentiles. It must be noted that, since these are single-year runs, impacts are likely to be underestimated since the model assumes reservoirs are at full storage at the beginning of the year. Moreover, the depth to the water table is the same across scenario runs. For these reasons a limited multiyear version of the model was created to allow these resources to dynamically vary to illustrate the benefits of the investment scenarios examined (discussed later in the section, "Long-Term Characteristics of Investments and Water Productivity").

Climate Risk Scenarios

To generate a wide range of potential climate scenarios, combinations of corresponding inflow and crop water requirement parameters are used. Inflow is varied from 10 to 90 percent exceedance probability using 10 percent increments, and the crop water requirement is varied from +2.5, +5, to +20 percent, corresponding to 1°–4.5°C temperature increases (possibly occurring around the 2020s and 2080s, respectively (based on the GCM outputs from chapter 4). Furthermore, since much of the waters in the system originate from the Upper Indus Basin (UIB) in the Himalaya, climate change impacts (using corresponding temperature and precipitation changes) on snow and ice in the UIB, and ultimately on the inflows into the Indus main-stem basin (as described in table 4.10) are included. From these, a total of 70 different climate futures are generated. These scenarios represent a plausible range of climate change futures within the next 80 years consistent with recent observations and theory. The impacts of these climate futures on the computable general equilibrium (CGE) and IBMR outputs are shown in figure 6.2.

Generally, negative impacts are estimated under these climate risk scenarios. Gross domestic product (GDP), Ag-GDP, and household income are estimated to decrease by 1.1, 5.1, and 2 percent, respectively. In the most extreme climate future (when inflow is 90 percent exceedance probability and the temperature increases +4.5°C), GDP, Ag-GDP, and household income are estimated to decrease by 2.7, 12, and 5.3 percent, respectively. Figure 6.3 also demonstrates that most of the negative impacts on incomes will occur for those households outside of the agriculture sector (except for those living in provinces other than Punjab and Sindh). Since the increase in prices is larger than the decrease in production, farm-related households will likely benefit. However, non-farm households (for example, urban) will have to pay more for food, thus resulting in decreasing household incomes. When the aggregated household income is calculated at the national level, the model weighted each household against their baseline incomes. Since non-farm households have higher weights, the aggregated household income shows a negative impact.

Figure 6.2 CGE and IBMR Outcomes under Climate Risk Scenarios

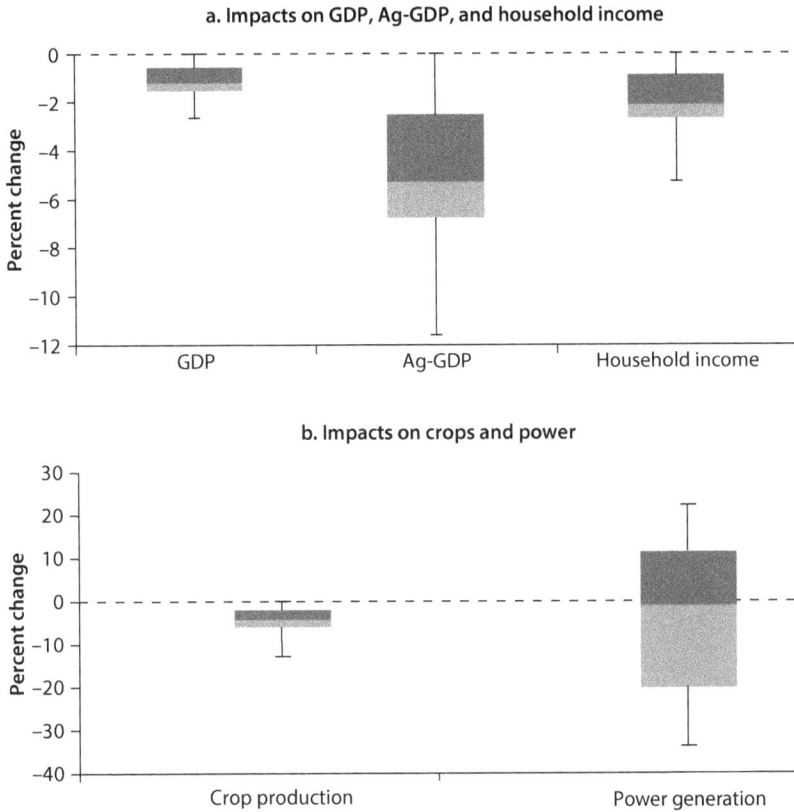

a. Impacts on GDP, Ag-GDP, and household income

b. Impacts on crops and power

Note: CGE = computable general equilibrium, IBMR = Indus Basin Model Revised. The upper error bar represents 99 percent, the upper box represents 75 percent, the middle line of the box represents 50 percent, the lower box represents 25 percent and the lower error represents 1 percent.

Figure 6.3 Household Income Changes under Various Climate Scenarios for Different Households

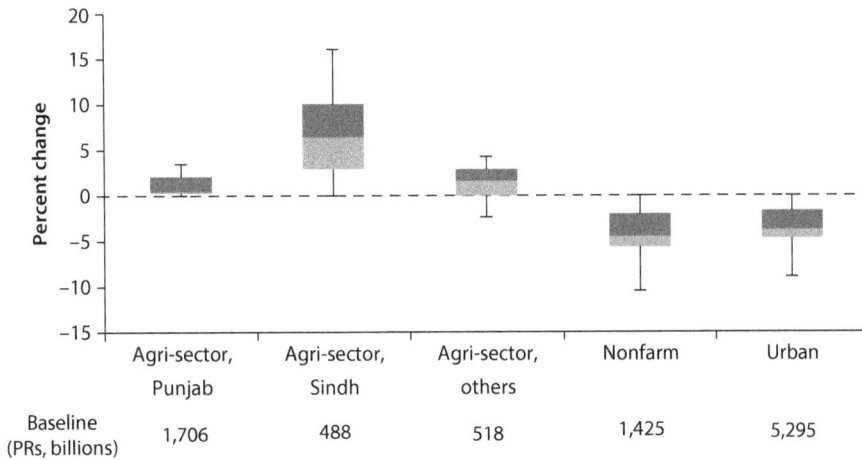

	Agri-sector, Punjab	Agri-sector, Sindh	Agri-sector, others	Nonfarm	Urban
Baseline (PRs, billions)	1,706	488	518	1,425	5,295

Figure 6.2b shows that total crop production is estimated to decrease 0–13 percent. The change in power generation varies the most, from +22 percent to –34 percent. Increases are due to more surface water becoming available from more snow-melt.

Figure 6.4 breaks down the crop production into different provinces and crops. The production changes are greatest in Sindh (around 10 percent on average). In the most extreme climate future, a 36 percent decrease in crop production is estimated in Sindh and a 5 percent decrease in Punjab. Figure 6.4b shows five crops that contribute most to the total crop revenue (see table 5.2). The largest projected production decrease will be for irrigated rice and sugarcane where, in the worst case scenario, almost 25 percent and 20 percent decreases, respectively, are estimated (6 percent and 5.7 percent average decrease). The worst-case scenarios for cotton and wheat are reductions of 2 percent and 7 percent, respectively. Basmati rice has a very small negative impact (less than 1 percent) under these climate futures. Note that these impacts do not consider changes in the biological crop yield response (beyond those changes due to water require-ments) in these simulations. According to Iqbal et al. (2009), for instance, using a bio-physiological based model, wheat yields are expected to decrease about 3 percent under the A2 scenario and 5 percent under B2 in the 2080s. Thus, these changes would be in addition to what this study's model currently predicts.

Figure 6.4 Crop Production Changes under Climate Risk Scenarios

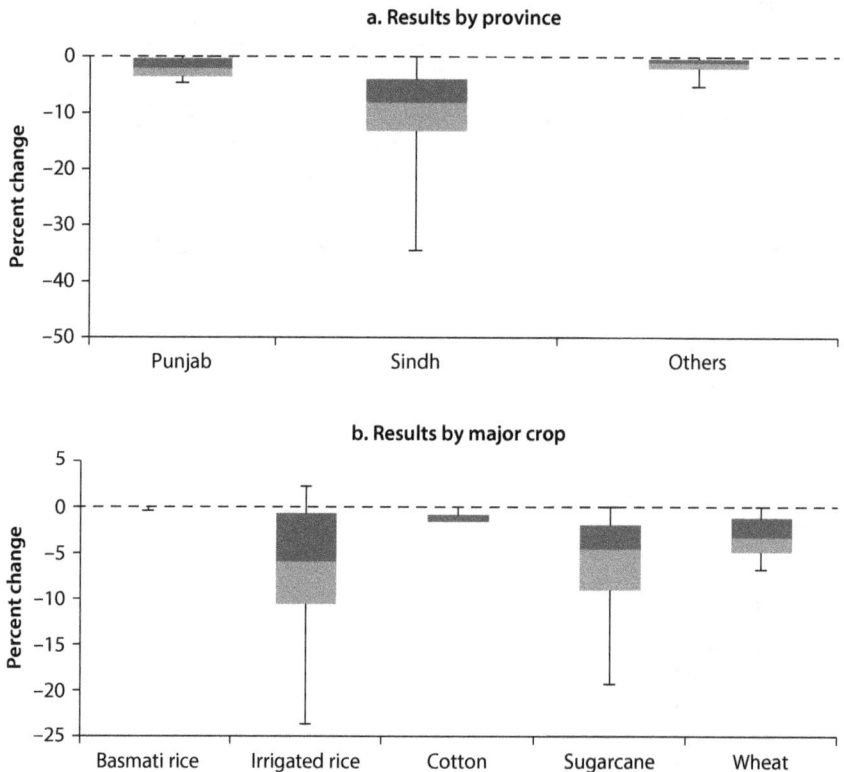

a. Results by province

b. Results by major crop

Figure 6.5 Commodities Revenue under Climate Risk Scenarios

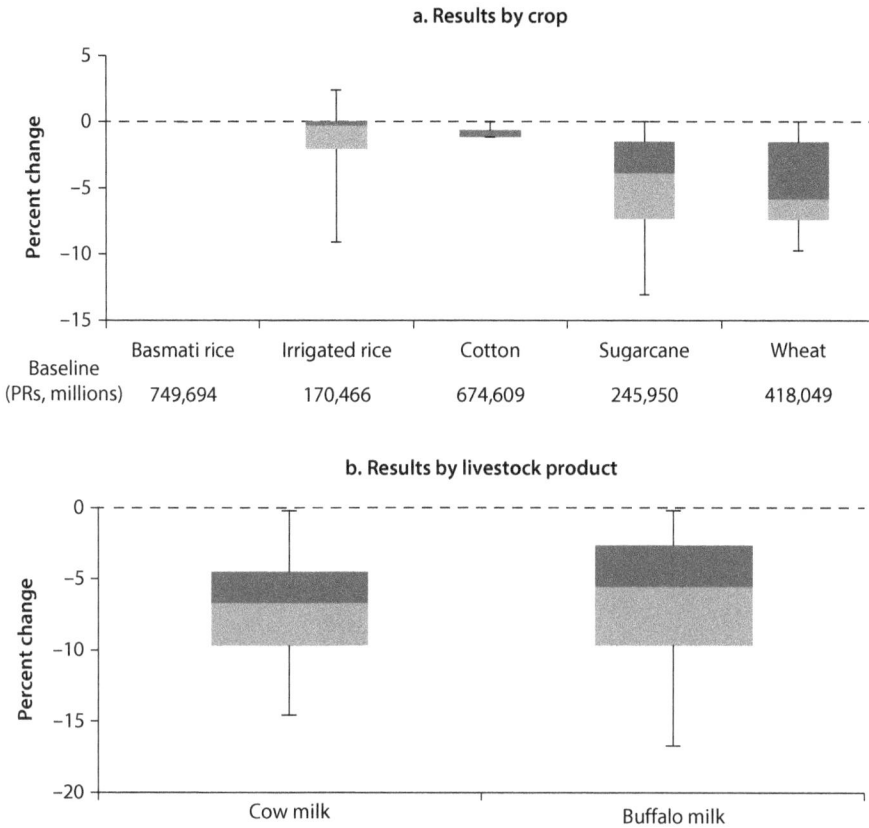

a. Results by crop

| Baseline (PRs, millions) | Basmati rice 749,694 | Irrigated rice 170,466 | Cotton 674,609 | Sugarcane 245,950 | Wheat 418,049 |

b. Results by livestock product

Figure 6.5 shows that the largest changes in revenue are estimated for wheat, sugarcane, irrigated rice, and cow and buffalo milk. In the extreme climate futures, sugarcane and irrigated rice revenues may drop by as much as 13 percent. Irrigated rice under the best circumstances may marginally increase in revenue.

Hydrograph Monthly Shift Scenario
The climate risk scenarios present the inflow, precipitation, and temperature change impacts under the assumption that the intra-annual hydrological pattern will remain the same. This section presents an evaluation of the effect of a shift of the hydrograph one month forward (April inflow becomes March inflow) and backward (April inflow becomes May inflow). A monthly shift forward is consistent with what a warming climate might do as described earlier (see chapter 4) that is, earlier snow melt and peak flow. Figure 6.6a shows that a forward monthly shift can have a larger negative impact on the economy than a backward shift. This impact is larger in magnitude than the average climate risk scenario. Figure 6.6b shows also the crop production and hydropower generation impacts. Less power is generated with these hydrograph shifts since less water is stored.

The Indus Basin of Pakistan • http://dx.doi.org/10.1596/978-0-8213-9874-6

Figure 6.6 CGE and IBMR Outcomes under Hydrograph Monthly Shift Scenarios

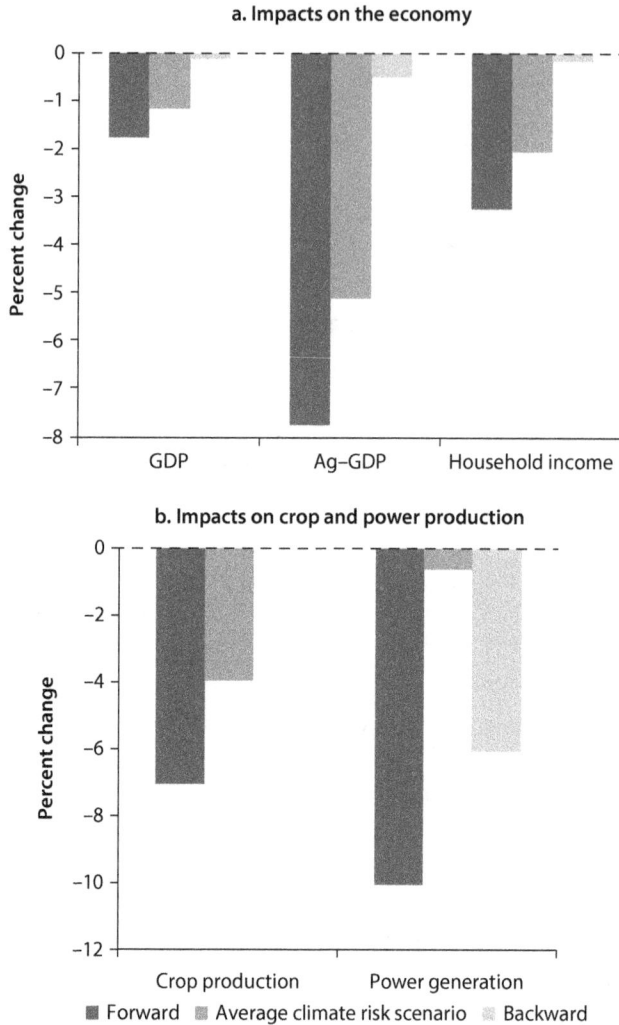

a. Impacts on the economy

b. Impacts on crop and power production

■ Forward   ■ Average climate risk scenario   ■ Backward

"Worst" and "Best" Case Climate Scenarios

The possible "worst" and "best" case climate futures represent low probability but possibly high impact scenarios ("surprise events"). The worst case is defined as the 90 percent exceedance probability inflow (only 10 percent of flows are less than this level), a forward monthly hydrograph shift, 20 percent less rainfall, 20 percent more water requirement (consistent with a +4.5°C change), and groundwater table depths 20 percent deeper throughout the basin. The best case is defined as the 10 percent exceedance probability inflow (90 percent of flows are less than this level), 20 percent more rainfall, no change in the existing crop water requirements, and groundwater table depths 20 percent shallower. Almost all GCM projections indicate increasing temperatures in the future and a high uncertainty in the direction and magnitude of precipitation change. However, given the counteracting nature that temperature and precipitation can play in overall water availability in the Indus system, the probability of the best and worst climate scenarios is believed to be quite small. That is not to say, however, that extreme events may increase in frequency in the future, a prediction whose science is yet inconclusive.

In the worst case, GDP, Ag-GDP, and household income decrease by 3.1, 13.3, and 6.1 percent, respectively, on an annual basis (figure 6.7a). In the best case, GDP, Ag-GDP, and household income increase by 1.0, 4.2, and 1.3 percent,

Figure 6.7 CGE and IBMR Outcomes under the "Worst" and "Best" Case Scenarios

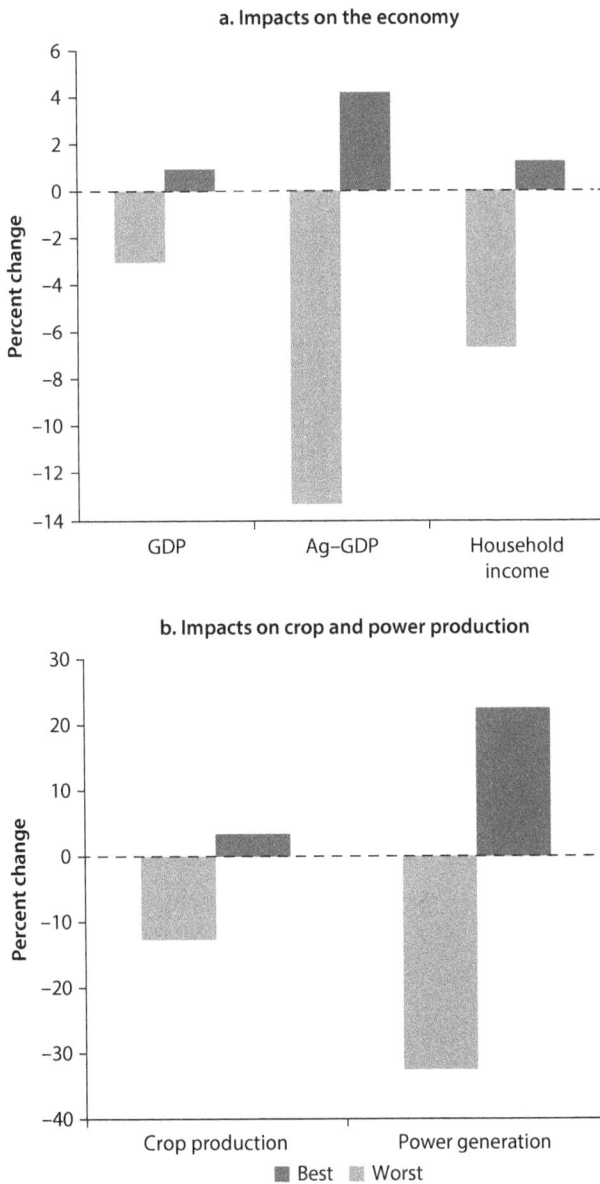

a. Impacts on the economy

b. Impacts on crop and power production

respectively. These ranges represent a range of possible economic futures in the basin. Figure 6.7b shows that the crop production decreases by 13 percent and increases by 3 percent under the worst and best scenarios, respectively. Similarly, power generation decreases by 32 percent and increases by 23 percent under worst and best scenarios, respectively. These results indicate that power generation is more sensitive to climate changes than crop production. This in part reflects the alternative source of irrigation (that is, groundwater) available for crop production and alternative cropping patterns.

Adaptation Investment Scenario Analyses

Three different investments are discussed in this section: canal and watercourse efficiency improvements (CANEFF), new reservoir construction, and crop yield improvement investments. To examine the role played by each of these investments over time the original IBMR is modified for a limited multiyear analysis (Indus Basin Multi-Year: IBMY). The IBMY uses the entire 50 year historical inflow time series (1961–2010). In the IBMY, the December reservoir storage from the previous year becomes the initial storage for the current year. Moreover, depth to water table is revised each year. All other data, including the crop water requirements, precipitation, price, and demand data, are assumed to be the same each year. Thus, only the water resource endowment is dynamic.

Improving System-Wide Efficiency

The first adaptation investment is to improve system-wide efficiency (CANEFF). The current canal and watercourse efficiency is estimated to be only 76 and 55 percent, respectively. Thus, only about 40–50 percent of the water in the system is actually available for field level irrigation. Several previous studies have addressed this issue (PRC Engineering 1986; World Bank 1996). For example, Cooley, Christian-Smith, and Gleick (2008) evaluated four different water-saving scenarios for the irrigation systems in California. Two commonly used technologies are sprinkler and drip/micro-irrigation systems. Sprinkler irrigation delivers water to the field through a pressurized pipe system and distributes it via rotating sprinkler heads, spray nozzles, or a single gun-type sprinkler. The field efficiency for sprinkler irrigation system is about 70–75 percent (Cooley, Christian-Smith, and Gleick 2008). Drip irrigation is the slow application of low-pressure water from plastic tubing placed near the plant's root zone. Drip systems commonly consist of buried PVC pipe mains and submains attached to surface polyethylene lateral lines. The field efficiency for sprinkler irrigation system is about 87.5–90 percent (Cooley, Christian-Smith, and Gleick 2008). Canal lining is another traditional approach to improving irrigation system efficiency. It can control seepage to save water for further extension of the irrigation network and also reduce waterlogging in adjacent areas (Swamee, Mishra, and Chahar 2000). Skogerboe et al. (1999) estimated that for the Fordwah Eastern Sadiqia project in Punjab, different types of canal lining can reduce the seepage losses by 50 percent. This study models an adaptation investment scenario whereby the system-wide efficiency is improved to 50 percent (from the existing 35 percent)—primarily through canal and watercourse improvements.

New Storage in the Indus Basin

The second adaptation investment is the construction of new reservoirs (NEWDAM). The construction of large dams can increase the country's water storage capacity and better manage scarcity. New dams will also add power generation; thereby helping to meet the country's expanding electricity needs. In this analysis, additional storage is primarily evaluated in terms of its ability to improve agricultural production for the existing irrigation system under climate change conditions. The potential economic value of storage for flood risk reduction, improved drought management, and expansion of the irrigated area is not included. Although hydropower production is estimated, the value of that electricity is not included here, and thus does not factor into GDP or objective function results. Thus, this evaluation should not be seen as a cost-benefit analysis of new dam construction. The adaptation investment used here introduces about 13 MAF into the modeling structure. The operation rules and storage-level relationship is assumed the same as the existing reservoirs in the system.

Improving Crop Technologies and Yields

The third adaptation investment is new crop technologies to improve crop yields (CYIELD). As noted earlier, it is assumed that crop yield is constant over time for each crop at each ACZ. However, it is reasonable to assume that crop yields will improve in the future as a result of new technologies and on-farm water management improvements (as shown in figure 2.10). For example, biotechnology investments in genetically modified (GM) crops promise great benefits for both producers and consumers of agricultural products, although the applications of GM are also associated with potential risks (FAO 2002).

One of the most successful examples of biotechnology is the application of BT cotton in China. Based on survey data, Huang et al. (2002) reported that farmers who used BT cotton observed increased output per hectare and increased their incomes due to reduced pesticides and labor inputs. Since no detailed GM crop data is available, a rough estimation was made of crop yield improvement based on FAO (2002) data on yield trends for different areas and different crops. For example, for developing countries the wheat yield improvement is about 2.0 percent per year and for rice is about 1.1 percent per year (over the 1989–99 period). The model includes an adaptation investment that assumes a 20 percent yield improvement, which will represent the possible yield in the next 10–20 years according to FAO estimates.

Long-Term Characteristics of Investments and Water Productivity

The cumulative distribution functions (CDF) of the IBMR objective value for the three investment scenarios just described from the 50-year historical record are presented in figure 6.8. The CDF is a graph that describes the probability of finding an objective value at that value or less. This 50-year simulation includes

Figure 6.8 Cumulative Distribution Functions of IBMR-2012 Objective Value for Different Adaptation Investments (without Climate Risk Scenarios)

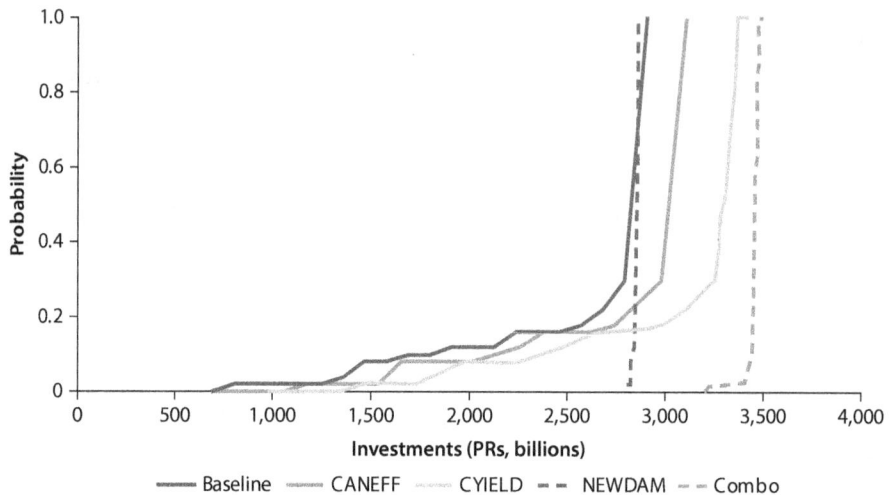

Note: CANEFF = canal and watercourse efficiency improvement, CYIELD = crop yield, NEWDAM = construction of new reservoirs. The cumulative distribution function (CDF) is a graph of the value of the objective function versus the probability that value will occur.

drought years and correspondingly low values of the objective value. The CANEFF and CYIELD investments shift the CDF to the right of the baseline, indicating that the average objective value tends to increase under these investments. The long left-side tails of the CDFs of these two investments are due to very low values that occurred in difficult years, such as the droughts that occur over the 50 year simulation. The NEWDAM investment is unique in that it eliminates the left-side tail, showing that additional storage reduces the probability of very low objective values, thus mitigating the effects of drought years. However, it does not increase the objective value under normal and high flow years. This is primarily because the objective function does not include the economic benefits from additional hydropower generation and flood control. In addition, while the increased reservoir volume may supply more water as a result of the constraints of the Accord, that water cannot be put to use effectively. "Combo" is all three investments combined. Summary statistics for these CDFs are shown in table 6.2. Notice that all mean values are higher than the baseline and that the standard deviation is reduced, especially so for the NEWDAM investment.

Performance of Adaptation Investment

This section is an evaluation of the performance of the adaptation investments under the range of future climate risk scenarios. The DIVACRD constraint is enforced in all model runs. Figure 6.9 and table 6.3 show that the CANEFF and CYIELD investments can significantly improve macroeconomic performance and household income under a climate change future. Instead of losses of 1, 5, and 2 percent for GDP, Ag-GDP, and household income, respectively, with these

Table 6.2 Mean and Standard Deviation Objective Value for 50 Years from Different Investments

Objective value (PRs, billions)	With DIVACRD	
	Mean	SD
Baseline	2,619	491
CANEFF	2,802	465
NEWDAM	2,843	38
CYIELD	3,085	466
Combo	3,451	34

Note: CANEFF = canal and watercourse efficiency improvement, CYIELD = crop yield, NEWDAM = construction of new reservoirs.

Figure 6.9 Economic Outcomes from CGE under Different Investments

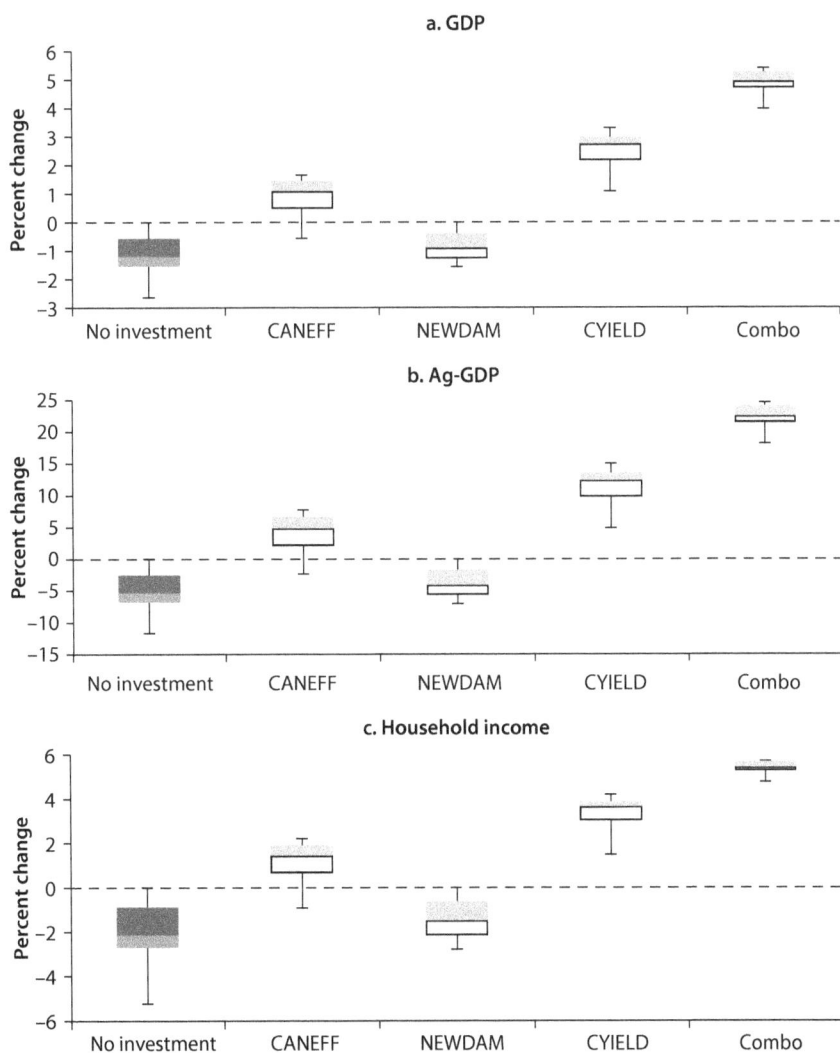

Note: CGE = computable general equilibrium, CANEFF = canal and watercourse efficiency improvement, CYIELD = crop yield, NEWDAM = construction of new reservoirs.

Table 6.3 Impact of Different Adaptation Investments under Climate Risks

	GDP	Agri-GDP	Household income
Average change without investments (%)			
No investment	−1.1	−5.1	−2
Average gain with investments (%)			
CANEFF	2.04	9.32	3.21
NEWDAM	0.29	1.5	0.64
CYIELD	3.66	16.7	5.42
Combo	6.05	27.4	7.45

Note: CANEFF = canal and watercourse efficiency improvement, CYIELD = crop yield, NEWDAM = construction of new reservoirs.

adaptation investments impacts are positive. For example, the average Ag-GDP will increase by about 4 percent and 11 percent with the CANEFF and CYIELD investments, respectively. The CANEFF and CYIELD investments show a clear positive shift with very low probabilities of observing negative changes. The NEWDAM investment shows minor improvement and reduces the impact of the 1st percentile climate future. As discussed in the previous section, this finding reflects primarily the contribution of additional storage to an existing irrigation system and does not incorporate other potential benefits to the economy and households.

Examining the impact of these investments on crop production and power generation (figure 6.10) shows that the relative efficacy of these investments on crop production is similar. The CANEFF and CYIELD investments result in greater crop production (5–11 percent more on average) than the NEWDAM investment. The NEWDAM investment, on the other hand, can minimize the impacts of extreme climate impact losses and reduce variability. Moreover, the power generation benefits can be quite large with the NEWDAM investment. The highest power generation increase is 130 percent. Even under the worst climate scenario power generation still increases by 20 percent. The economic value of new reservoirs under this analysis would be almost entirely from these power benefits and from a reduction in the impacts of extreme events.

Investment Costs

The cost of a system-wide canal efficiency program (to achieve the 50 percent scenario) is difficult to quantify because of different approaches used and diversity in geographic conditions. Skogerboe et al. (1999) report that the cost for canal lining in Pakistan (for the Fordwah Eastern Sadiqia Project) ranged PRs 608–3,725 per foot of canal (in 1999 PRs). The reduction in seepage ranged from a factor of 2–10, depending on the prevalent conditions. Using these indicative costs estimates and assuming that all 60,000 km of watercourses in the IBIS are lined, a cost range of PRs 180–1,107 billion is calculated. Similarly, system-wide efficiency can be improved directly at the on-farm levels. Cooley, Christian-Smith, and Gleick (2008) estimated the unit cost for sprinkler (US$1,000–3,500 per acre) and for drip/micro-irrigation systems

Figure 6.10 IBMR Outcomes under Different Investments

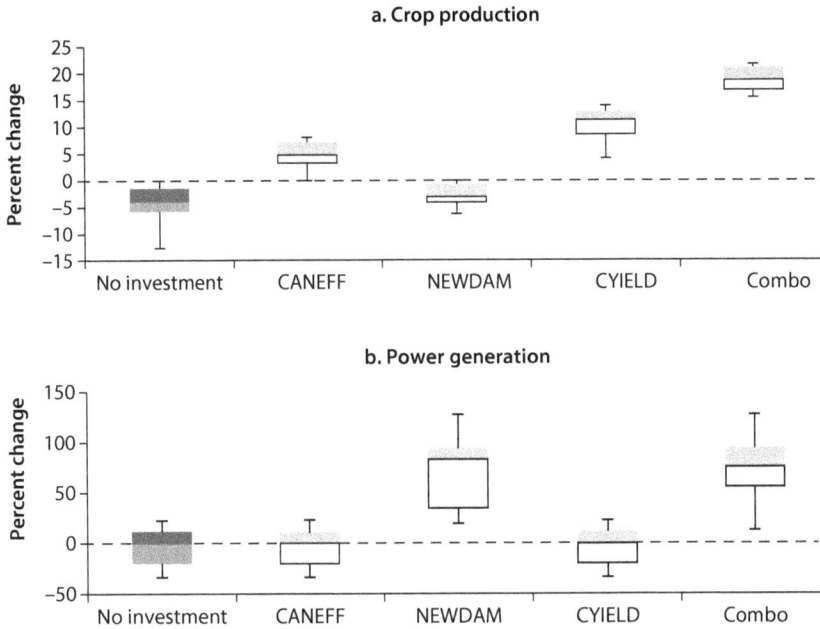

a. Crop production

b. Power generation

Note: IBMR = Indus Basin Model Revised, CANEFF = canal and watercourse efficiency improvement, NEWDAM = construction of new reservoirs, CYIELD = crop yield.

(US$500–2,000 per acre). The cost for new storage can be found on the Water and Power Development Authority (WAPDA) website. The estimated cost for Basha and Kalabagh (two often discussed reservoirs) is PRs 390,000 million and PRs 366,000 million in 2005 value, respectively. Therefore, the total estimated cost for new reservoirs construction is PRs 390,000 + 366,000 = PRs 756 billion (US$9.54 billion). Finally, estimating the investment cost required for new technologies and research and development to raise crop yields by 20 percent is difficult because of the inherent complexity associated with these investments. According to a study by Menrad, Gabriel, and Gylling (2009), the additional costs per tons for GM and non–genetically-modified rapeseed oil, sugar, and wheat are PRs 32,400, 20,160, and 15,680, respectively, in Germany. Therefore, the average additional cost per tons is PRs 22,746.

Effect of Investments on Food Self-Supply

Having access to sufficient quantities of food is an indicator of food security. Thus, changes in crop production are directly related to Pakistan's ability to be food self-sufficient, notwithstanding the role that food imports may play. Wheat-based products (flour and bread) are a major part of the diet in Pakistan. These provide upwards of 60 percent of the protein and carbohydrate in the average Pakistani diet (Bastin, Sarwar, and Kazmi 2008). Supply and demand of wheat are used to estimate the impact that climate change may have on the nutritional requirements in Pakistan. Bastin, Sarwar, and Kazmi (2008) calculated a conversion factor from wheat production to combined protein and carbohydrate supply in flour. The average value is 70 percent, which means that 1,000 tons of wheat can provide 700 tons of protein and carbohydrate in flour. This converting factor is then multiplied by the 50-year average wheat production and the protein and carbohydrate supply are calculated under baseline and all adaptation investment scenarios. Using the GCM projections from chapter 4, six temperature and precipitation combinations were selected for the 2020s, 2050s, and 2080s (see table 6.4).

The protein and carbohydrate requirement is estimated based on the population. Historical data show a strong linear relationship between Pakistan total population and total requirement of protein and carbohydrate. This equation is used to estimate the protein and carbohydrate requirement in 2020s, 2050s, and 2080s, based on future population estimates. The results of protein and carbohydrate supply and requirement in all these years under all investment scenarios are given in table 6.5. The supply is higher than the demand during the baseline and 2020s time period. However, the supply will be less than demand by the 2050s without any investment. Only the CYIELD investment can maintain the production to meet the future protein and carbohydrate requirements.

Table 6.4 Projected Temperature, Precipitation, and Inflow Changes

Years	Projected temperature (°C)	Projected precipitation (%)	Crop water requirement change (%)	Inflow change (%)
2020s	+1.5	Low: 0	+4	−4
		High: +10	+4	+4
2050s	+3	Low: −10	+10	−8
		High: +20	+10	+17
2080s	+4.5	Low: −10	+20	+1
		High: +10	+20	+18

Note: Temperature and precipitation projection follow the average general circulation model (GCM) results in chapter 4; inflow changes from current condition are calculated by the model in chapter 3.

Table 6.5 Protein and Carbohydrate Supply and Requirements under Climate Change Estimates

Pakistan population (millions)		Cereal-based protein and carbohydrate demand (tons, millions)	Protein and carbohydrate supply (tons, millions)			
			Baseline	CANEFF	NEWDAM	CYIELD
Baseline	167.4	10.1	16.3	18.0	16.4	19.8
2020-low P	227.8	13.7	16.1	17.7	16.2	19.4
2020-high P			16.2	17.8	16.3	19.5
2050-low P	307.2	18.4	15.8	17.2	15.9	19.0
2050-high P			15.9	17.4	15.9	19.1
2080-low P	386.7	23.1	15.5	16.8	15.6	18.6
2080-high P			15.5	16.8	15.6	18.7

Note: CANEFF = canal and watercourse efficiency improvement, CYIELD = crop yield, NEWDAM = construction of new reservoirs. Shaded cells mean supply is less than demand.

By the 2080s, none of the investments can supply sufficient protein and carbohydrates for the country. Disaggregating these findings by province shows that Punjab is able to meet its protein and carbohydrate demands, even out to 2080. The real food security challenge will be in Sindh, even as early as 2020. Note that it is assumed that interprovincial trading does not change and that food imports are not considered.

Environmental Issues

The primary environmental issues related to water use in the Indus Basin include flow requirements to the sea, groundwater over pumping, and groundwater salinity. These issues were discussed and analyzed in Ahmad and Kutcher (1992). These outcomes are reevaluated here under the climate risk scenarios and investment scenarios described earlier, and the cost for a sustainable groundwater usage situation is also evaluated.

Current Environmental Conditions

Environmental flows to the sea are required to sustain the diverse deltaic ecosystems and minimize saline intrusion. A minimum 10 MAF to the sea is required per the 1991 Provincial Accord. This minimum flow is difficult to maintain during drought years (for example, 2002–04). Haq and Khan (2010) estimate that over the last 20 years, at least 2 million acres of arable land have been lost in Sindh as a result of salt water intrusion. On average, over the long-term historical record, almost 30 MAF is available to the sea (figure 6.11). This, however, may be an issue in the future if current trends continue. Figure 6.12 shows that the flows below Kotri Barrage (the last barrage in the system) have decreased over time. The annual average from 1936 to 1960 was 87 MAF compared to 41 MAF over the 1977–2000 time period. For future analysis, this modeling

Figure 6.11 Multiyear Flow to the Sea, 1961–2009

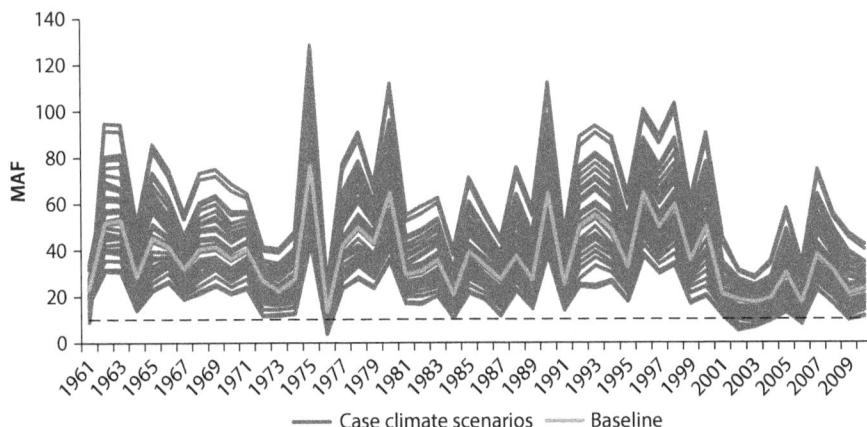

Note: MAF = million acre-feet.

Figure 6.12 Historical Flows below Kotri Barrage, 1938–2004

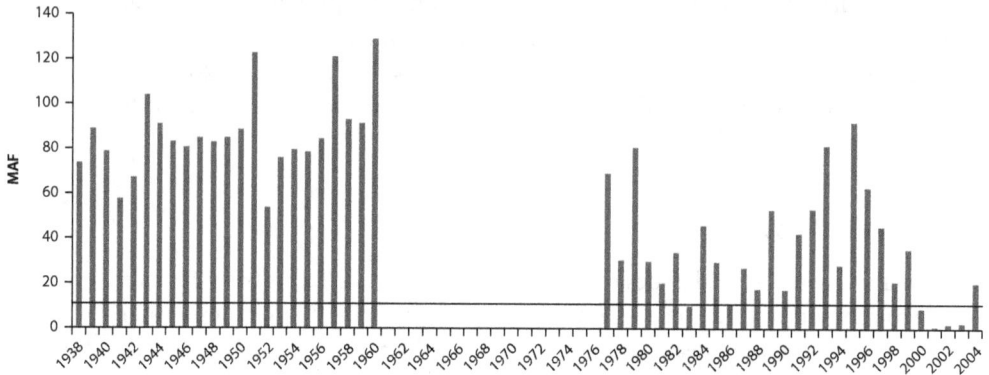

Note: Solid line represents 10 million acre-feet (MAF) established under the 1991 Provincial Water Allocation Accord.

Table 6.6 Baseline Environmental Conditions

	Area (acres, millions)		Net recharge (MAF)		Salt balance in soil layer (tons, millions)	
Province	Fresh	Saline	Fresh	Saline	Fresh	Saline
Punjab	18.1	4.8	−9.6	4.4	+35.9	+4.2
Sindh	3.4	9.4	2.7	4.6	+5.4	+29.3
Others	3.0	0.4	−2.5	0.2	+4.5	+0.7

Note: MAF = million acre-feet.

framework can be used to test the system-wide implications of various scientifically-based monthly minimum flow requirements.

Groundwater quantity and quality issues are also a prominent environmental issue in the Indus Basin. Punjab faces unsustainable pumping rates while in Sindh the dominant issue is related to problems of salinity and waterlogging. On average, the net recharge in freshwater areas in Punjab (groundwater inflow minus outflow) is –9.6 MAF and thus the water table is declining (1–6 ft per year). This situation is worst during drought years. On average, the net recharge in the saline areas in Sindh is +4.6 MAF (more groundwater is flowing in than out); as a result the net salt accumulation on the surface in these areas is more than 29 million tons (per year). The reason for this net accumulation is that groundwater pumping does not exist in these saline areas, so fresh water is not recharged into these aquifers. Meanwhile, the evaporation rates in saline areas are usually higher than in fresh areas (due to the higher water table), generating substantial quantities of salt near the ground surface (root zone). The net accumulation of salt in the fresh water areas in Punjab is also quite large because of the large volumes of water being applied for irrigation (which have some background salinity). The baseline groundwater conditions are given in table 6.6.

Sustainable Groundwater Usage

To evaluate the sustainability of current groundwater usage, assuming an energy cost of PRs 5 per kwh (WAPDA) and pumping depths around 80 ft (this depth

Table 6.7 Baseline vs. Sustainable Groundwater Usage Model

	Baseline	Capped pumping	Difference	Percentage of baseline
Objective value (PRs, billions)	2544	2506	38	98
Total production (1,000 tons)	94,047	89,385	−4,662	95
Punjab	64,983	61,428	−3,555	95
Sindh	24,225	23,434	−791	97
Canal diversion (MAF)	109.6	109.5	−0.1	..
Punjab	58.1	58.3	0.2	101
Sindh	43.3	42.8	−0.5	99
Groundwater pumping (MAF)	57.9	50.0	−7.9	86
Punjab	54.1	47.0	−7.1	87
Sindh	3.2	2.6	−0.6	82

Note: .. = negligible, MAF = million acre-feet.

to groundwater is the deepest value anywhere in the model), the total pumping cost used in the model is PRs 800 per AF for the baseline case. The minimum groundwater required is about 4.8 MAF where farmers have little choice but to pump from the aquifer, even when the pumping costs are very high. Because groundwater is always needed to augment surface irrigation supplies, the reduction in groundwater pumping also serves as a cap on productivity and, consequently, surface water use. The total pumping in which the net recharge is zero (that is no drop in water table) is calculated in the model to be about 48.6 MAF. This may be considered as the "safe yield" and matches earlier reported numbers (for example, 51 MAF by Qureshi 2011).

Table 6.7 shows the results of restricting the model to a groundwater abstraction at the safe yield of about 50 MAF. The table shows the economic cost for sustainable groundwater usage. The objective value decreases by PRs 38 billion (US$0.47 billion). This represents only a 2 percent reduction, which suggests that prudent policy on groundwater management may be cost-effective, depending on an assessment of resource values. Punjab will have the most impacts in terms of crop production. On the other hand, these "costs" of sustainable groundwater usage will be more equal between Punjab and Sindh when the provincial allocation constraint is relaxed. The pumping reductions are greatest in Punjab. Note that the actual depth to groundwater does not directly affect the optimization. Thus, these results may be optimistic.

Climate and Investment Scenarios

This section presents changes in these environmental parameters under different adaptation investments. Figure 6.13 shows the results of flow to the sea and fresh groundwater net recharge with and without adaptation investments. The flow to the sea does not significantly change when adaptation investments are introduced. Part of this can be explained by examining how much surface and groundwater is used (table 6.8). The CYIELD investment uses almost the same amount of surface water as the no investment scenario. Thus, the remaining flow

Figure 6.13 Environmental Related Outcome under Different Investments

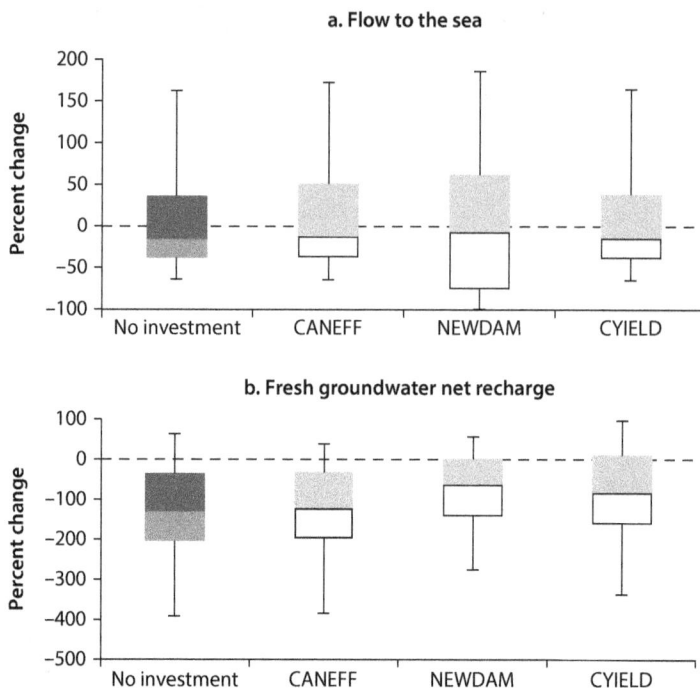

a. Flow to the sea

b. Fresh groundwater net recharge

Note: CANEFF = canal and watercourse efficiency improvements, CYIELD = crop yield, NEWDAM = construction of new reservoirs.

Table 6.8 Irrigation Mix under Different Adaptation Investments

Average water uses under climate risk scenarios (MAF)	Canal	Tubewell	Total
No investment	109.8	66.4	176.2
CANEFF	108.1	51.6	159.7
NEWDAM	115.4	64.2	179.6
CYIELD	109.5	62.6	172.1

Note: MAF = million acre-feet, CANEFF = canal and watercourse efficiency improvements, CYIELD = crop yield, NEWDAM = construction of new reservoirs.

to the sea is essentially the same. The CANEFF investment, on the other hand, diverts less surface water during the high and normal flow situations but almost the same amount during the low flow situations. This is because canal water is "free" compared to groundwater. Therefore, when CANEFF makes more surface water available, the system will divert the same amount of surface water and dramatically reduce the groundwater usage (since groundwater has a cost). Since canal diversions are almost the same, the flow to the sea value will also be the same. The range of values of flow to sea for the NEWDAM investment increases. During high flow situations, the 1991 Provincial Accord limits the amount of water that can be diverted and utilized. Thus, additional water provided by new storage cannot be used and escapes to the sea (since the model does not allow for the expansion of irrigated areas). During low flow situations, the additional storage will allow the system to divert more water from canal and result in less flow to the sea. This is the reason for a wider range in NEWDAM.

Similarly, the groundwater net recharge does not change significantly with these investments. Only under the NEWDAM investment is the groundwater net recharge improved. This is because more water is made available for the surface system, particularly during drought conditions. Water losses from canals and watercourses are treated as the major groundwater inflow in the model. When efficiency improves (as in the CANEFF investment), the amount of canal diversion decreases and the losses also decrease. This is a negative effect in groundwater net recharge. Moreover, pumping is reduced, which is a positive effect in groundwater net recharge (table 6.8). Thus, these two effects offset each other. With the CYIELD investment, slightly less groundwater is used and net recharge marginally improves.

Figure 6.14 shows the salt balance in both fresh and saline areas resulting from the adaptation investments. In this study, we follow the approach taken by

Figure 6.14 Salt Accumulation in Soil Layer in Fresh and Saline Areas

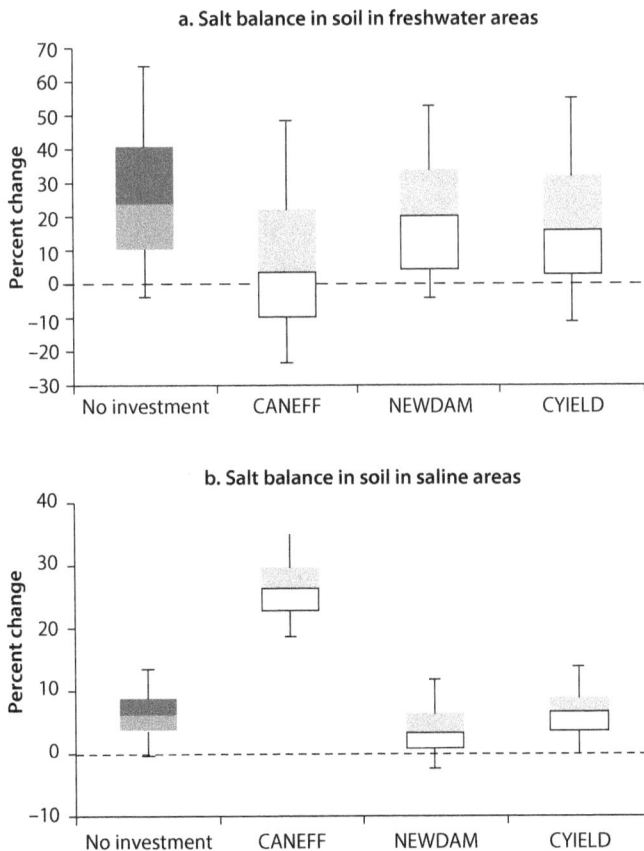

a. Salt balance in soil in freshwater areas

b. Salt balance in soil in saline areas

Note: CANEFF = canal and watercourse efficiency improvements, CYIELD = crop yield, NEWDAM = construction of new reservoirs.

Ahmad and Kutcher (1992) in these calculations. For fresh areas, the largest component of salt accumulation is from pumping groundwater with background salinity. Only CANEFF, which reduces overall groundwater use, can reduce the salt accumulation in these fresh areas. With reduced fresh water flows into the saline areas (under CANEFF), the salt concentrations increase. NEWDAM and CYIELD have no significant effect on salt balance in either fresh or saline areas.

In summary, this chapter has examined the potential agro-economic impacts of some of the pressing challenges introduced in chapters 1 and 2. It has also assessed potential investment and management alternatives for the Indus Basin. This analysis was accomplished by linking an updated IBMR model with an economically broader CGE–social accounting matrix (SAM) analysis. The analysis first identified some of the key sensitivities and more robust aspects of the IBIS. At every step, key data and modeling issues, and further Indus irrigation management questions were encountered, which will be priorities for further analysis. The concluding chapter presents these main findings, their significance, and priorities for future research.

References

Ahmad, M., A. Brooke, and G. P. Kutcher. 1990. *Guide to the Indus Basin Model Revised.* Washington, DC: World Bank.

Ahmad, M., and G. P. Kutcher. 1992. "Irrigation Planning with Environmental Considerations: A Case Study of Pakistan's Indus Basin." World Bank Technical Paper 166, World Bank, Washington, DC.

Bastin, G. Q., S. Sarwar, and Z. A. Kazmi. 2008. "Wheat-Flour Industry in Pakistan." Discussion Paper, Competitiveness Support Fund, Islamabad.

Briscoe, J., and U. Qamar, eds. 2006. *Pakistan's Water Economy: Running Dry* (includes CD of background papers, 2005). Washington, DC: World Bank.

Brown, C., and J. W. Hansen. 2008. *Agricultural Water Management and Climate Risk: Report to the Bill and Melinda Gates Foundation.* IRI Techical report 08-01, International Research Institute for Climate and Society, Palisades, NY.

Cooley, H., J. Christian-Smith, and P. H. Gleick. 2008. *More with Less: Agricultural Water Conservation and Efficiency in California—A Special Focus on the Delta."* Pacific Institute, Oakland, CA.

FAO (Food and Agriculture Organization of the United Nations). 2002. *World Agriculture: Towards 2015/2030 Summary Report.* Rome.

Haq, N. U., and M. N. Khan. 2010. *Pakistan's Water Concerns.* IPRI Factfile, Islamabad Policy Research Institute, Islambad. http://ipripak.org/factfiles/ff127.pdf (accessed May 17, 2012).

Huang, J., R. Hu, C. Fan, C. E. Pray, and S. Rozelle. 2002. "Bt Cotton Benefit, Costs and Impact in China." *AgBioForum* 5 (4): 153–66.

Immerzeel, W. W., F. Pellicciottti, A. Gobiet, and S. Ragetti. 2011. "The Impact of Uncertainty in Climate Change Scenarios on Projection of Future Water Supply from the Asian Water Towers." *Eos Trans. AGU* 92 (52) Fall Meeting. Suppl., Abstract C53E-0721.

Iqbal, M. M., M. A. Goheer, S. A. Noor, H. Sultana, K. M. Salik, and A. M. Khan. 2009. *Climate Change and Wheat Production in Pakistan: Calibration, Validation and Application of CERES-Wheat Model*. GCISC-RR-14, Global Change Impact Studies Centre, Islamabad.

Menrad, K., A. Gabriel, and M. Gylling. 2009. "Costs of Co-Existence and Traceability Systems in the Food Industry in Germany and Denmark." Paper presented at the Fourth International Conference on Coexistence between Genetically Modified and non-GM based Agricultural Supply Chains, Melbourne, Australia, November 10–12.

Michel, A. A. 1967. *The Indus Rivers: A Study of the Effects of Partition*. New Haven, CT: Yale University Press.

Naheed, G., and G. Rasul. 2010. "Projections of Crop Water Requirement in Pakistan under Global Warming." *Pakistan Journal of Meteorology* 7 (13): 45–51.

PRC Engineering. 1986. *Technical Criteria for Rehabilitation of Canal Systems in Pakistan*. Islamabad: Government of Pakistan/USAID Irrigation Systems Management Project.

Qureshi, A. S. 2011. "Water Management in the Indus Basin in Pakistan: Challenges and Opportunities." *Mountain Research and Development* 31 (3): 252–60.

Skogerboe, G. W., M. Aslam, M. A. Khan, K. Mahmood, S. Mahmood, and A. H. Khan. 1999. *Inflow-Outflow Channel Losses and Canal Inning Cost-Effectiveness in the Fordwah Eastern Sadiqia (South) Irrigation and Drainage Project*. Report R-85, International Water Management Institute, Lahore, Pakistan.

Swamee, P., G. C. Mishra, and B. Chahar. 2000. "Minimum Cost Design of Lined Canal Sections." *Water Resources Management* 12: 1–12.

Tariq, S. M., and S. Ul Mulk. 2005. "Sustainable, Accountable Institutions." Background paper 7 for *Pakistan's Water Economy: Running Dry*. Washington, DC: World Bank and Oxford University Press.

WAPDA (Water and Power Development Authority). 1990. *Water Sector Investment Planning Study (WSIPS)*. 5 vols. Lahore, Pakistan: Government of Pakistan Water and Power Development Authority.

World Bank. 1996. *Pakistan Impact Evaluation Report*. Report 15863-PAK, World Bank, Washington, DC.

CHAPTER 7

The Way Forward

The years 2009–11 offer a perspective on the current challenges of water and agriculture, along with mounting future uncertainties Pakistan's government faces at both federal and provincial levels. From the dramatic increase in food prices to the floods of 2010 (resulting in US$10 billion in damages and losses), managing these many forms of resource variability and uncertainty has become vital for this agricultural and water economy. The prospects of climate change may amplify these concerns. Moreover, with continued growing populations and increasing water demand across all sectors, these risks must be anticipated and managed.

This study focuses on the impacts of climate risks and various alternatives faced by water and agriculture managers on water and crop productions in the Indus Basin of Pakistan. This study analyzes the inter-relationships among the climate, water, and agriculture sectors to support the prioritization and planning of future investments in these sectors. The important role that water management plays in the productivity of the agriculture sector and overall food security is recognized in many different forums and policy reports in Pakistan. However, these linkages are not always comprehensively addressed (with systems-based models) in federal and provincial planning documents and budgeting. This study provides a systems modeling framework for these purposes.

Several different models are integrated in this work. This includes a model of Upper Indus snow and ice hydrology (critical for determining overall water availability), an optimization agro-economic model of the Indus Basin Irrigation System (IBIS), and an updated computable general equilibrium (CGE) model of Pakistan's wider macro-economy. This integration of models helps to frame recommendations for strengthening water, climate, and food security planning, policies, and research priorities for the Indus Basin.

This analysis first identified some of the key hydroclimatic sensitivities and more robust aspects of the IBIS. Second, the models used here are among the best mathematical representations available of the physical and economic responses to these exogenous future climate risks. However, like all modeling approaches, uncertainty exists as parameters may not be known with precision and functional forms may not be fully accurate. Thus, careful sensitivity analysis and an understanding and appreciation of the limitations of these models are required. Further collection and analysis of critical input and output observations (for example, snow and ice data) will enhance this integrated framework methodology and future climate impact assessments.

Key Messages

- *Given the orographic complexity of the Upper Indus Basin, future climate impacts are highly uncertain.* The variability of the system is moderated by snow and ice dynamics. Using a simple model of the dynamics of the water and energy balance in the Upper Indus Basin, glacier melt contributes an estimated 15 percent of flow to the total. That is, the system is largely a snowmelt-dominated basin in which future changes will largely be dependent on changes in winter precipitation. Thus, future changes are difficult to predict because of counterbalancing effects of changes in temperatures and precipitation. Because of the complexities at these high elevations, general circulation models (GCMs) are unlikely to have much value for forecasting purposes. There is a need for major investment in snow and ice hydrology monitoring stations, further scientific research, and forecasting to improve the hydrologic predictability of the Upper Indus Basin.

- *The 1991 Provincial Water Allocation Accord (DIVACRD) is a critical hydrologic constraint in the system.* According to model results, if the Accord is relaxed and optimal economic allocation between and within provinces is allowed, both Punjab and Sindh could benefit. The system-wide net revenue will increase by about PRs 158 billion (almost US$2 billion): PRs 83 billion additional in Punjab and PRs 82 billion additional in Sindh. Moreover, by relaxing the Accord and implementing an economically based water allocation mechanism, provinces will be better able to manage extreme events by more reliably meeting system-wide demands. This would need to be supported by investment in effective, transparent, real-time water delivery measurement systems; capacity-building in Indus River System Authority (IRSA) and Water and Power Development Authority (WAPDA) for technical decision-support systems and forecasting; and equally substantial investment in trust-building among stakeholders. Even though it is unlikely and probably unwise that the Accord constraint should—*by itself*—be relaxed, there is room for flexible policy adjustments and mechanisms within the wider framework of the present Accord (for example, interprovincial exchange of surplus allocations, water banking, and leasing arrangements), which the modeling results suggest should be pursued on agro-economic grounds.

- *Future climate risks are estimated to impact the macro-economy and households.* Gross domestic product (GDP), Ag-GDP, and household income are estimated to decrease on average in the future by 1.1, 5.1, and 2.0 percent, respectively. In the most extreme future (when inflow is 90 percent exceedance probability and the temperature increases +4.5°C), GDP, Ag-GDP, and household income are estimated to decrease by 2.7, 12, and 5.5 percent, respectively.

- *Future climate change risks could impact crops in Sindh more than in Punjab.* Sugarcane, wheat, cotton, and irrigated rice demonstrate the greatest climate sensitivity. The model results indicate that under the most extreme climate condition, crop production in Sindh is estimated to decrease by 36 percent, compared to 5 percent in Punjab. Sugarcane will be affected the most both in terms of production (20 percent) and revenue (13 percent). Furthermore, these impacts are conservative as the model used the assumption that the biological crop yield response is unchanged due to the heat stress, disease, and extreme events caused by climate risk (beyond those changes due to water requirement). To be effective, especially in the context of changes under the 18th amendment to its constitution, Pakistan's new climate change policy will require substantial advances in coordination among federal, provincial, and local agencies—and new mechanisms for the transparent real-time information exchange with private-sector and civil society organizations.

- *Non-farm household incomes will suffer more under climate risk scenarios due to increased crop prices.* In general, crop production is estimated to decrease and crop prices increase. Since the increase in prices is larger than the decrease in production, farm-related households will likely benefit. However, non-farm households (for example, urban) will have to pay more for food, thus resulting in decreasing household incomes.

- *An increase in basin-wide storage will increase the hydropower generation and minimize the impacts of extreme events.* However, additional storage will not significantly increase agricultural benefit (assuming no expansion of the current irrigated area). A modeled storage-yield relationship is determined and demonstrates that in terms of food production, additional storage will not have a significant influence, especially for high flow or normal flow years. Only during the drought years can additional storage maintain food production at a normal level and probably only for a year and not for multiyear sustained events. The frequency of future extremes is inconclusive. The main economic benefits from storage will be from power generation. This finding recommends that a near-term modeling priority be the integration of energy security, both in the power sector, which includes hydel, and the petroleum sector, which includes light diesel oil used by the majority of tubewells in Pakistan.

- *Different adaptation investments show potential to minimize the impacts of future climate risks and meet food security objectives.* Investments in canal and watercourse efficiency and in crop technologies are estimated to increase average crop production by 5–11 percent. This will have positive impacts on

The Indus Basin of Pakistan • http://dx.doi.org/10.1596/978-0-8213-9874-6

the macro-economy and households. These investments are still vulnerable under low-flow drought conditions. On the other hand, investment in additional storage (NEWDAM) significantly reduces these impacts and this interannual variability.

- *Climate change is estimated to impact future food availability on a nationwide scale.* The supply of protein and carbohydrates from wheat will be lower than the forecast future demand under climate change by the 2050s and 2080s. Model results suggest that the investment in crop technologies to improve yields can help to balance this. Disaggregating these findings by province shows that Punjab is able to meet its protein and carbohydrates demands, even out to the 2080s. The real challenge will be in Sindh, even as early as the 2020s, under the assumption that interprovincial patterns do not change. This concern is amplified by the regionally low level of current food security in Sindh in the 2011 National Nutrition Survey.

- *Groundwater depletion in the fresh water area and basin-wide salinity issues will become worse if no policy intervention is made.* The analysis reveals that groundwater is a key resource in the Indus and that the net revenues lost (as a percentage of the baseline) are not that significant under a scenario where safe groundwater yields are enforced. However, the long-term trends are troubling. The net recharge in fresh groundwater areas are negative in all provinces with the largest values estimated in Punjab, which suggests continued declining water tables. This also contributes to increased saline water intrusion. In addition, salt accumulation is positive in all provinces and in both fresh water and saline areas. Given the scale of these issues, a new phase of truly visionary planning is needed for conjunctive management of surface and groundwater management.

Final Thoughts

The precise impact of these climate risks on the Indus Basin remains to be seen. This much is known, however: climate change will pose additional risks to Pakistan's efforts to meet its water and food security goals, which are key to reducing poverty, promoting livelihoods, and developing sustainably. As the Pakistan population grows, the ability to meet basic food requirements and effectively manage water resources will be critical for sustaining long-term economic growth and rectifying widespread food insecurity and nutrition deficiencies. These are challenges above and beyond what Pakistan is already currently facing, as evidenced during the 2009–11 time period. Strategic prioritization and improved planning and management of existing assets and budget resources are critical. These strategic choices will be largely dependent on a sound assessment of the economics of these impacts.

The integrated systems framework used in this analysis provides a unique, broad approach to estimating the hydrologic and crop impacts of climate change risks, assessing the macro-economic and household-level responses, and developing an effective method for assessing a variety of adaptation investments and policies. In assessing the impacts, several different modeling environments must be integrated to provide a more nuanced and complete picture of how water and food security interrelate. Moreover, such a framework allows for extensive scenario analysis to identify and understand key sensitivities. This is critical to making decisions in a highly uncertain future. Finally, through this integration of multiple disciplines, a richer and more robust set of adaptation investment options and policies for the agriculture and water sectors can be identified and tested. Continued refinements to the assessment approach developed in this volume will further help to sharpen critical policies and interventions by the Pakistan Government.

Structure of IBMR

Description of 12 Agro-Climatic Zones

North West Kabul Swat (NWKS)

This agro-climatic zone (ACZ) was originally named NWFP in the previous model setting but was renamed NWKS in the current version. The water for this ACZ is not fed from the Indus irrigation system. It diverts water from the Kabul, Swat, and Warsak Rivers before the water reaches the Indus. Water shortages can largely be traced to limited canal capacities. Groundwater is usable throughout most of the zone; cropping is dominated by sugarcane, maize, and wheat.

North West Mixed Wheat (NWMW)

This ACZ originally belonged to PMW (described below) in the previous model setting. It has been separated out since it is geographically located in NWFP province. The water for this ACZ is fed from a canal, the Chashma Right Bank. The primary crop is wheat, with some rice and sugarcane.

Punjab Mixed Wheat (PMW)

This ACZ on the left bank of the Indus below Jinnah barrage contains nearly 2 million culturable command areas (CCA). The topography is rough, soils are sandy, and seepage is high, resulting in low cropping intensities and yields. The dominant crops are wheat, rice, cotton, and sugarcane.

Punjab Rice Wheat (PRW)

This ACZ contains 2.8 million CCA, virtually all of which is underlain by fresh groundwater. This has spurred intense private tubewell development. As a result, cropping intensities are among the highest in Punjab, with Basmati rice being the dominant cash crop. Relatively high returns to farming, combined with a shortage of labor have led to rapid mechanization.

Punjab Sugarcane Wheat (PSW)

This ACZ lies between PMW and PRW, and contains about 4.4 million CCA. Wheat and sugarcane are the principal crops. About one-third of the zone is saline, but farmers make extensive use of groundwater in fresh areas. Water shortages that do exist are largely attributable to low watercourse efficiencies.

Punjab Cotton Wheat West (PCWW) and Punjab Cotton Wheat East (PCWE)
Originally, PCW was the largest ACZ in the model, comprising over 11 million CCA on the left bank of the Indus. It is split between west and east in its current version. PCWW is fed from the Upper Indus Canals, and PCWE is fed from the Jhelum River. The main crops, cotton and wheat, have some of the highest yields in Pakistan. About one-fourth of the ACZ suffers from salinity. Groundwater is extensively used in the rest of the zone, but providing adequate water remains an overall constraint.

Sindh Cotton Wheat North (SCWN) and Sindh Cotton Wheat South (SCWS)
These two ACZs cover 3 million CCA each. Nearly half of the north and most of the south are saline or waterlogged or both. Yields from areas remaining in use are favorable. Groundwater use is minimal, and surface water supplies are hampered by high losses, particularly at the watercourse level.

Sindh Rice Wheat North (SRWN) and Sindh Rice Wheat South (SRWS)
These two ACZs are the right and left bank delineations of the Sindh wheat and rice zone. About two-thirds of the 4.4 million CCA in the north are saline and the entire south is similarly classified. Because of the high water table, yields for other crops are poor, and cropping intensities, particularly in the south, are lowest in the Basin. Surface water supplies are adequate, although other inputs, such as fertilizer are used sparingly.

Balochistan Rice Wheat (BRW)
This ACZ originally belonged to SRWN in the previous model setting. It has been separated since it is geographically located in Balochistan province. The water for this ACZ is fed from the Pat Feeder and Kirther canals. The primary crops are rice, wheat, some onion, and sugarcane.

Table A.1 Hierarchal Structure of Provinces, Agro-Climatic Zones, and Canal Command Areas

Provinces (4)	ACZs (12)	ACZ name	Canals and subarea definition (%)[a] (49)
NWFP	NWMW	North_west_mixed_wheat	22-USW.S1 (100), 22A-PHL.S1 (100)
			23-LSW.S1 (100), 24-WAR.S1 (100)
			25-KAB.S1 (100)
	NWKS	North_west_kabul_swat	27-CRB.S1 (100)
Punjab	PMW	Punjab_mixed_wheat	26-THA.S1 (35), 26-THA.S2 (17)
			26-THA.S3 (30), 26-THA.S4 (18)
			26A-GTC.S1 (86), 26A-GTC.S2 (14)
			28-MUZ.S1 (25)
	PCWW	Punjab_cotton_wheat_west	20-PAN.S1 (70), 20-PAN.S2 (30)
			21-ABB.S1 (100), 28-MUZ.S2 (75)
			29-DGK.S1 (100)
	PCWE	Punjab_cotton_wheat_east	01-UD.S1 (100), 02-CBD.S2 (50)
			06-SAD.S1 (100), 07-FOR.S1 (100)
			08-PAK.S1 (100), 09-LD.S1 (100)
			10-LBD.S1 (50), 10-LBD.S2 (50)
			15-BAH.S1 (80), 15-BAH.S2 (20)
			16-MAI.S1 (65), 16-MAI.S2 (35)
			17-SID.S1 (100), 19-RAN.S1 (100)
	PSW	Punjab_sugarcane_wheat	11-JHA.S2 (49), 11-JHA.S3 (19)
			12-GUG.S2 (53), 12-GUG.S3 (21)
			13-UJ.S1 (100), 14-LJ.S1 (64)
			14-LJ.S2 (36), 18-HAV.S1 (100)
	PRW	Punjab_rice_wheat	02-CBD.S1 (50), 03-RAY.S1 (100)
			04-UC.S1 (100), 05-MR.S1 (100)
			11-JHA.S1 (32), 12-GUG.S1 (26)
Sindh	SCWN	Sindh_cotton_wheat_north	33-GHO.S1 (50), 33-GHO.S2 (50)
			37-KW.S1 (100), 38-KE.S1 (100)
			39-ROH.S1 (39), 39-ROH.S2 (20)
			41-NAR.S1 (20), 41A-RAI.S1 (100)
	SCWS	Sindh_cotton_wheat_south	39-ROH.S3 (16), 39-ROH.S4 (25)
			41-NAR.S2 (80)
	SRWN	Sindh_rice_wheat_north	31-DES.S1 (100), 32-BEG.S1 (50)
			32-BEG.S2 (50), 34-NW.S1 (100)
			35-RIC.S1 (100), 36-DAD.S1 (100)
	SRWS	Sindh_rice_wheat_south	42-KAL.S1 (100), 43-LCH.S1 (100)
			44-FUL.S1 (100), 45-PIN.S1 (100)
Balochistan	BRW	Balochistan_rice_wheat	30-PAT.S1 (100), 31A-KAC.S1 (100)
			34A-KIR.S1 (100)

Note: ACZ = agro-climatic zone.
a. Sub-area definition (%) represents the percentage of water this specific canal will deliver to this specific ACZ from its total intakes.

IBMR Updating to IBMR 2008

The last version of Indus Basin Model Revised (IBMR) is based on data from 2000 (primarily the Agricultural Statistics of Pakistan and water-related data from the Water and Power Development Authority [WAPDA]) and earlier farm surveys (for example, 1976 XAES Survey of Irrigation Agriculture and the Farm Re-Survey in 1988 as part of the Water Sector Investment Planning Study [WSIPS]).

Model Structure Change: The Lateral Groundwater Flow in IBMR

Almost every aspect of hydrogeology is considered in the original IBMR structure except for lateral flow. In this study, this lateral flow was defined as the underground flow in or out of an agro-climatic zone (ACZ) due to the groundwater hydrologic gradient. Since there was no basic survey data available, this value is estimated.

Assuming that the lateral flow has a linear relationship with the change in water table, the following equation is considered:

$$D_{\text{lateral}} = kd2 \times (\Delta GD - kd1) \qquad (B.1)$$

where D_{lateral} is the lateral flow, ΔGD is the monthly water table change, and $kd1$ and $kd2$ are coefficients. A pre-defined IBMR simulation was set up to solve for $kd1$ and $kd2$ for each ACZ. The purpose of this predefined simulation is to search for a set of $kd1$ and $kd2$ for different ACZs and groundwater types that makes the groundwater depth at the end of the simulation match the long-term observed value. This procedure means that the lateral flow should balance the water flow in and out of the aquifer. Since the groundwater balance is a post-calculation after the optimization in the current structure, all of the economic outcomes from IBMR will not be affected by adding $kd1$ and $kd2$.

Model Structure Change: The Refined Sugar and Sugar Cane Issue

In the original IBMR (Ahmad, Brooke, and Kutcher 1990), sugarcane will produce two different end products: SC-GUR, which is treated as refined sugar and consumed at the farm level, and SC-MILL, which is the production of sugarcane that goes into the market. SC-MILL was redefined in IBMR and the refined sugar demand for the Indus River was modeled. This section describes the details of this modification.

Using the *Pakistan Sugar Annual 2009 Gain Report* (USDA 2009), the basin-wide production, demand, and price of both sugarcane and refined sugar are available. The model uses the price and demand of refined sugar to build the demand function in IBMR. Therefore, when the model optimizes the production, it will optimize the refined sugar production. However, the cropped area and the straw yield should be computed from sugarcane. A conversion coefficient between refined sugar and sugarcane is used to achieve this purpose. The value used in the model is 0.0865, which is described in Ahmad, Brooke, and Kutcher (1990) and is also similar to the value reported by the USDA (2009). The relationship between sugarcane and refined sugar production (unit as weight) is:

$$\text{Refined Sugar} = 0.0865 \times \text{Sugarcane} \tag{B.2}$$

This coefficient is used to adjust the yield from sugarcane to refined sugar both as the unit of weight per area of land and also the straw conversion coefficient for SC-MILL. Meanwhile, since the model computes refined sugar production, SC-MILL was added as one of the consumable crops. The on-farm consumption ratio of refined sugar mentioned in Ahmad, Brooke, and Kutcher (1990) was used as the refined sugar demand for 2008–09.

Model Structure Change: Removed Variables and Equations in IBMR 2008

The tractor and private tubewell numbers in the model are considered appropriate, so further investment in tractors and private tubewells is not necessary and was therefore removed from the model. The related constraints are also removed. Draft power is 99 percent provided by tractors in Pakistan. Therefore, the provision for draft power from bullocks is removed from the model. The removal of the bullock requirement is problematic, since it is one of the meat sources in the model. The fixed-cost of bullock is much higher than cow. Under this circumstance, bullock will never be raised. Therefore, the bullocks-cow population constraint is changed to force the model to maintain a certain amount of bullocks in each ACZ. In addition, one item is added to describe the tractor cost by multiplying the price of tractor per hour per acre with the tractor power requirement of different crops, months, and ACZs.

Data Updating

Price Update

The crop prices from the "Agricultural Statistics of Pakistan (ASP) 2008–2009" (Government of Pakistan, Ministry of Food, Agriculture and Livestock 2010) are collected by region. A simple mapping check was first conducted to assign the ACZs in IBMR into the nearest region. The ratio from 2000 and 2008 ASP prices is used to update the crop price data for IBMR. The livestock prices (milk and meat) are updated with a similar procedure. Wages, protein cost, tractor cost, tubewell cost, seed cost, water cost and all other miscellaneous cost are updated based on the change in gross domestic product (GDP). A rate of 1.51 is used to update all the mentioned prices for IBMR.

Demand Update

The demand data are used to construct the demand curves in IBMR. Since the crop price has been updated, the demand should also be updated, based on the assumption that the slope of the demand curve will remain the same in 2000 and 2008. The new demands are then back calculated by fixing the slope of demand curve with the given 2008 price.

On-Farm Consumption Update

The on-farm consumption should also be updated for the new baseline. According to Ahmad, Brooke, and Kutcher (1990), the values of on-farm consumption are computed by multiplying the data in estimated total production by the proportions of produce consumed on the farm from the re-survey. Following this concept, an updated total production table for 2008 is computed first using data in the 2008 ASP. The on-farm consumption ratio is assumed to remain the same from 2000 to 2008. Using the same ratio the on-farm consumption is updated at ACZ level.

Yield Update

The observed crop yield data from the 2008 ASP were used to compare the 2000 observed yield with the IBMR 2000 baseline. Most of the crops have similar values allowing the national crop yields to be directly updated to IBMR baseline using ASP 2008 values. However, some crops—cotton, gram, orchard, and fodder—have either larger differences or have no data for updating. Therefore, the cotton yield (seed cotton) for the IBMR baseline is updated using Food and Agricultural Organization of the United Nations data. And for all other missing crops, yields are calculated using the average ratio of 2008/2000 ASP crop yields multiplied by the IBMR 2000 baseline crop yield values.

IBMR Model Diagnosis

Since IBMR is an optimization model, the results cannot be expected to match uniquely the observed values. Therefore, we do not try to validate the model with observation but rather diagnose the model to check if the crop production and area shows a similar pattern as the observed. The purpose is to understand the

performance of the model under baseline conditions as well as the difference
between observations and the baseline. The primary outputs of IBMR are
agricultural products; therefore, the factors checked are cropped area, crop
production, and livestock production.

The observation data are all summarized from the 2008–09 ASP, and the
comparisons have been done at the provincial level. Punjab and Sindh are two
major provinces that rely on the irrigated network from the Indus River. The
IBMR shows better results in the cropped area and production for these two
provinces, as shown in figures B.1 and B.2. Almost all crops are at the same

Figure B.1   Cropped Area and Production from IBMR Baseline and ASP 2008–09 in Punjab

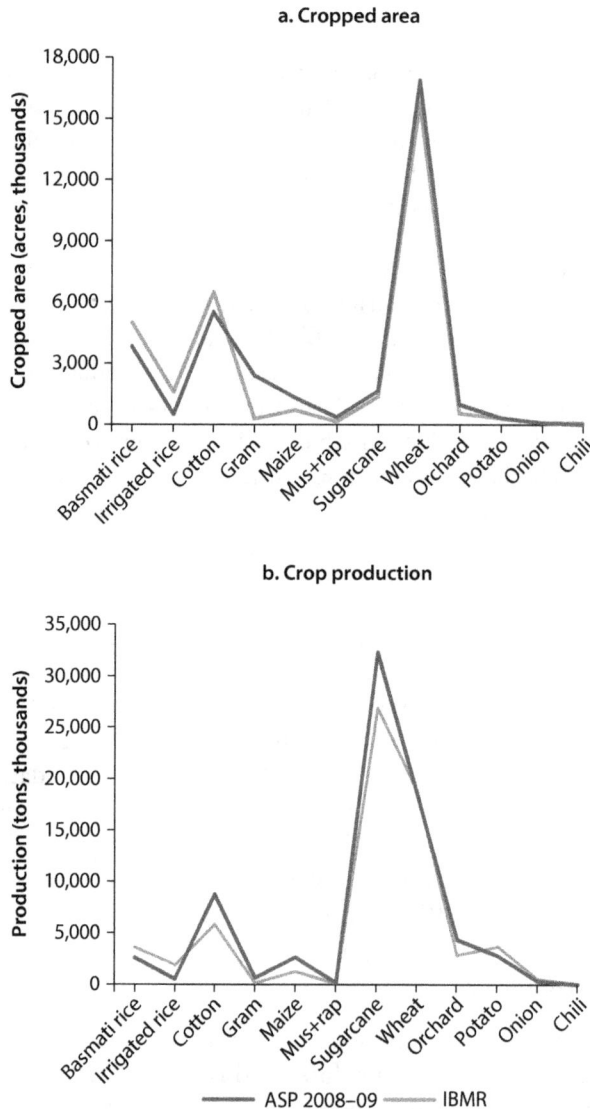

a. Cropped area

b. Crop production

Note: IBMR = Indus Basin Model Revised, ASP = Agricultural Statistics of Pakistan.

magnitude for the modeling result and the observation except for SC-MILL. In IBMR, we used the parameters (price, yield and consumption) of refined sugar to model this commodity. A possible reason for the underestimation might be due to the price underestimate and also government subsidies on sugarcane. The R^2 for cropped areas are 0.98 and 0.98 for Punjab and Sindh, respectively. And the R^2 for production are 0.99 and 0.99 for Punjab and Sindh, respectively. These results show that the model captures the trend of cropped area and production very well. Although the absolute values might be different, the relative cropped

Figure B.2 Cropped Area and Production from IBMR Baseline and ASP 2008–09 in Sindh

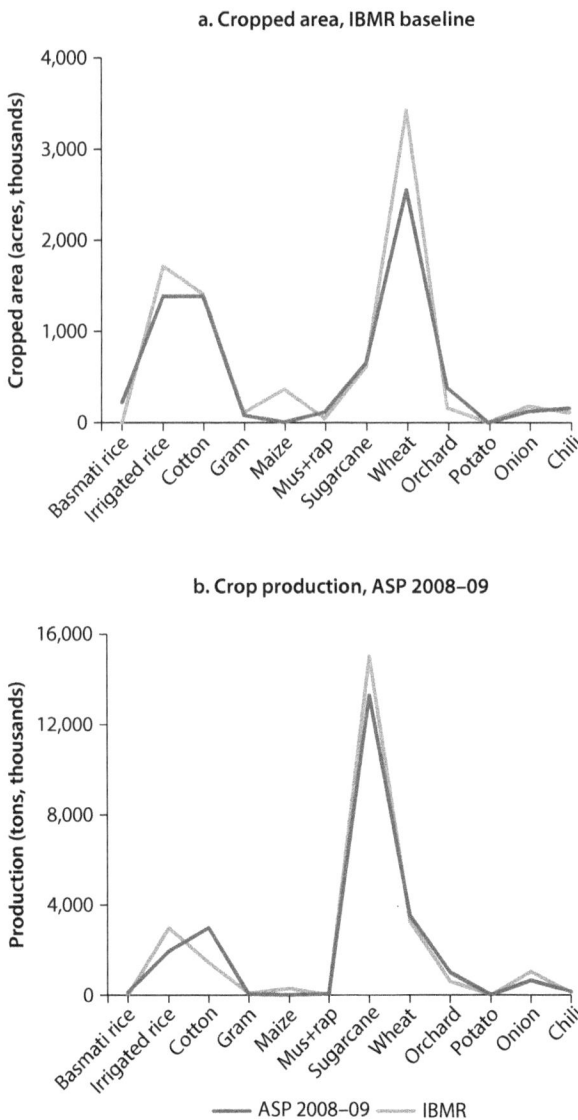

a. Cropped area, IBMR baseline

b. Crop production, ASP 2008–09

Note: IBMR = Indus Basin Model Revised, ASP = Agricultural Statistics of Pakistan.

The Indus Basin of Pakistan • http://dx.doi.org/10.1596/978-0-8213-9874-6

pattern (which means the proportion of each crop in area and production) is very similar to reality.

Figures B.3 and B.4 show the cropped area and crop production in NWFP (North-West Frontier Province) and Balochistan, respectively. The modeling results are underestimated in these two provinces, which can be expected because only the irrigated area was modeled and only small portions of these two provinces are covered by the irrigated network in reality. Ahmad, Brooke, and Kutcher (1990) used a coefficient of determination (R^2) to test if the model can

Figure B.3 Cropped Area and Production from IBMR Baseline and ASP 2008–09 in NWFP

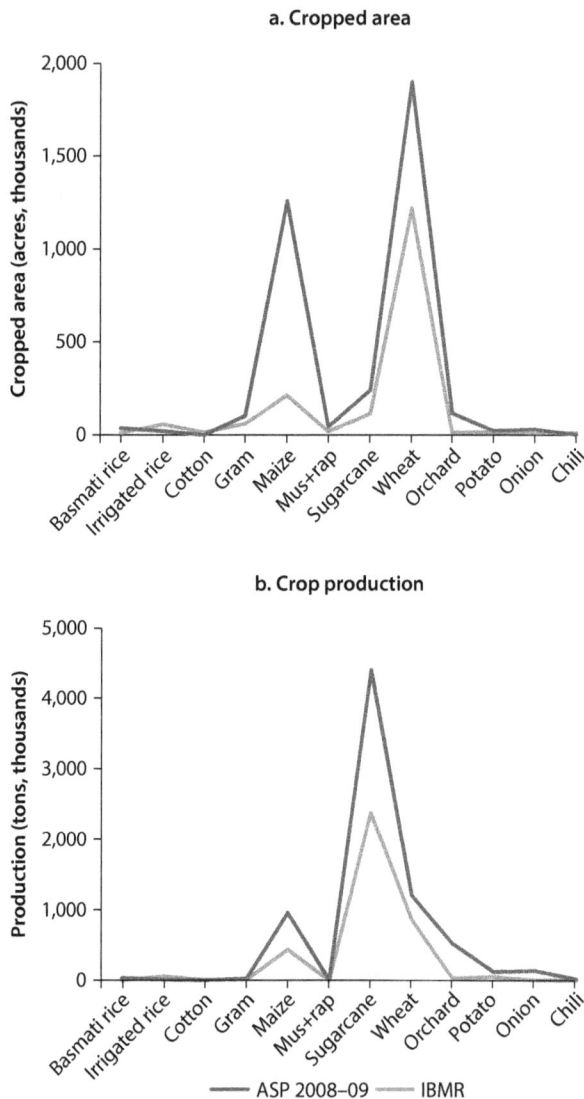

a. Cropped area

b. Crop production

Note: IBMR = Indus Basin Model Revised, ASP = Agricultural Statistics of Pakistan, NWFP = North-West Frontier Province.

at least capture the trend of cropped area and production. The R^2 for cropped areas are 0.90 and 0.83 for NWFP and Balochistan, respectively. And the R^2 for crop production are 0.98 and 0.41 for NWFP and Balochistan, respectively. Balochistan shows the largest differences between modeling results and observation. But since Balochistan only represents a very small portion of the entire Indus River, the results will not significantly affect the basinwide outcome.

Table B.1 shows the results of modeling livestock numbers compared to the ASP data. (Only Sindh province has 2008–09 data available, other provinces

Figure B.4 Cropped Area and Production from IBMR Baseline and ASP 2008–09 in Balochistan

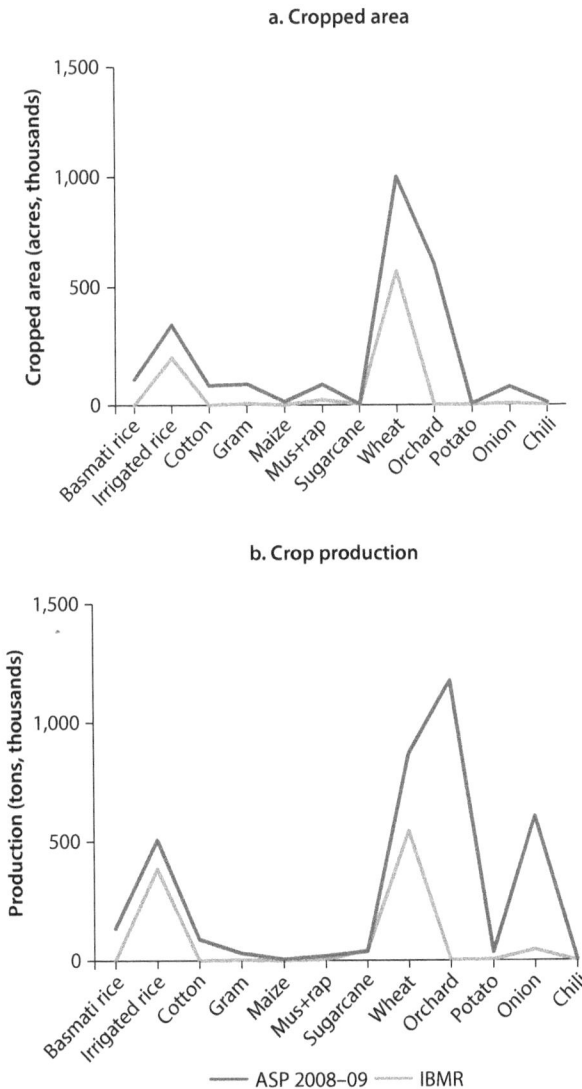

a. Cropped area

b. Crop production

Note: IBMR = Indus Basin Model Revised, ASP = Agricultural Statistics of Pakistan.

Table B.1 Livestock Comparison between IBMR 2008 and ASP 2008–09
animals, thousands

		Cow	Bullocks	Cattle	Buffalo
NWFP	IBMR	94	47	141	429
	ASP 2006			5968	1928
Punjab	IBMR	3074	1537	4611	6178
	ASP 2006			14412	17747
Sindh	IBMR	1602	801	2404	4117
	ASP 2008–09			6925	7340
Balochistan	IBMR	104	52	156	0
	ASP 2008			2254	320

Note: IBMR = Indus Basin Model Revised, ASP = Agricultural Statistics of Pakistan, NWFP = North-West Frontier Province.

have only 2006 data.) Generally, livestock is underestimated. A major reason is that much of the livestock production is in the non-irrigated areas and is not modeled in IBMR. Thus, the modeled livestock numbers are only half of the ASP data. The other possible reason is that in the model bullocks were removed and the ratio between cow and bullock was fixed. This modification might also affect the number of cows and bullocks in the model since the bullock population is capped by the cow population.

This appendix provides some diagnosis of the IBMR baseline run compared to observed data. Although the modeling results do not perfectly match the ASP data, they still provide a reasonable framework to assess the irrigated agro-activities in the Indus River Basin.

References

Ahmad, M., A. Brooke, and G. P. Kutcher. 1990. *Guide to the Indus Basin Model Revised.* Washington, DC: World Bank.

Government of Pakistan, Ministry of Food, Agriculture, and Livestock. 2010. "Agricultural Statistics of Pakistan 2008–2009." Islamabad, Pakistan.

USDA (U.S. Department of Agriculture). 2009. "Pakistan Sugar Annual 2009 Gain Report." USDA Foreign Agricultural Service-PK9005, U.S. Department of Agriculture, Washington, DC.

Details of the Pakistan Social Accounting Matrix (SAM)

Table C.1 SAM Accounts

A-WHTI1	Wheat irrigated Sindh	A-HORT3	Fruits/vegetables Other Pakistan	A-TRWAT	Transport-Water
A-WHTN1	Wheat non-irrigated Sindh			A-TRAIR	Transport-Air
A-PADI1	Rice IRRI (irr) Sindh	A-FOR3	Forestry Other Pakistan	A-TROTH	Transport-Other
A-PADB1	Rice basmati (irr) Sindh	A-CATT	Livestock (cattle, milk)	A-CONSH	Construction and housing
A-COTT1	Cotton (irr) Sindh	A-POUL	Livestock (poultry)	A-BSERV	Business Services
A-CANE1	Sugar cane (irr) Sindh	A-FISH	Fishing	A-ESERV	Education
A-OCRP1	Other field crops Sindh	A-MINE	Mining	A-HSERV	Health care
A-HORT1	Fruits/vegetables Sindh	A-VEGO	Veg Oils	A-PERSV	Personal Services
A-FOR1	Forestry Sindh	A-WHTF	Wheat Milling	A-OSERV	Other Priv Services
A-WHTI2	Wheat irrigated Punjab	A-RICI	Rice Milling (Irri)	A-PUBS	Public Services
A-WHTN2	Wheat non-irrigated Punjab	A-RICB	Rice Milling (Bas)	C-WHT	Wheat
A-PADI2	Rice IRRI (irr) Punjab	A-SUG	Sugar	C-PADI	Unmilled Rice IRRI (irr)
A-PADB2	Rice basmati (irr) Punjab	A-OTHF	Other food	C-PADB	Unmilled Rice basmati (irr)
A-COTT2	Cotton (irr) Punjab	A-LINT	Cotton gin (lint)	C-COTT	Cotton (irr)
A-CANE2	Sugar cane (irr) Punjab	A-YARN	Cotton spin (yarn)	C-CANE	Sugar cane (irr)
A-OCRP2	Other field crops Punjab	A-CLTH	Cotton weave (cloth)	C-OCRP	Other field crops
A-HORT2	Fruits/vegetables Punjab	A-KNIT	Knitwear	C-HORT	Fruits/vegetables
A-FOR2	Forestry Punjab	A-GARM	Garments	C-CATT	Livestock (cattle, milk)
A-WHTI3	Wheat irrigated Other Pakistan	A-OTXT	Oth Textiles	C-POUL	Livestock (poultry)
		A-WOODL	Wood and leather	C-FOR	Forestry
A-WHTN3	Wheat non-irrigated Other Pakistan	A-CHEM	Chemicals	C-FISH	Fishing
		A-CEM	Cement, bricks	C-MINE	Mining
A-PADI3	Rice IRRI (irr) Other Pakistan	A-PETR	Petroleum refining	C-VEGO	Veg Oils
A-PADB3	Rice basmati (irr) Other Pakistan	A-MANF	Other Manufacturing	C-WHTF	Wheat Milling
		A-ENRG	Energy	C-RICI	Milled IRRI Rice
A-COTT3	Cotton (irr) Other Pakistan	A-TRADW	Trade-wholesale	C-RICB	Milled Basmati Rice
A-CANE3	Sugar cane (irr) Other Pakistan	A-TRADR	Trade-retail	C-SUG	Sugar
		A-TRADO	Trade-other	C-OTHF	Other food
A-OCRP3	Other field crops Other Pakistan	A-TPTLAN	Transport-Land	C-LINT	Cotton Lint

table continues next page

Table C.1 SAM Accounts *(continued)*

C-YARN	Cotton yarn	LA-MF1	Labor-agric (own)-med Sindh	H-LF1	Large farm Sindh
C-CLTH	Cotton Cloth			H-LF2	Large farm Punjab
C-KNIT	Knitware	LA-MF2	Labor-agric (own)-med Punjab	H-LF3	Large farm Other
C-GARM	Garments			H-MF1	Med farm Sindh
C-OTXT	Other Textiles	LA-MF3	Labor-agric (own)-med OPak	H-MF2	Med farm Punjab
C-LEAT	Leather			H-MF3	Med farm OthPak
C-WOOD	Wood	LA-SF1	Labor-agric (own)-sm Sindh	H-SF1	Small farm Sindh
C-CHEM	Chemicals	LA-SF2	Labor-agric (own)-sm Punjab	H-SF2	Small farm Punjab
C-CEM	Cement, bricks			H-SF3	Small farm OthPak
C-PETR	Petroleum	LA-SF3	Labor-agric (own)-sm OPak	H-0F1	Landless Farmer Sindh
C-MANF	Other Manufacturing	LA-AGW	Labor-agric (wage)	H-0F2	Landless Farmer Punjab
C-ENRG	Energy	LA-SKU	Labor-non-ag (unsk)	H-0F3	Landless Farmer OthPak
C-CONS	Construction	LA-SK	Labor-non-ag (skilled)	H-AGW1	Waged rural landless farmers Sindh
C-TRADW	Wholesale Trade	LN-LG1	Land-large-Sindh		
C-TRADR	Retail Trade	LN-LG2	Land-large-Punjab	H-AGW2	Waged rural landless farmers Punjab
C-TRADO	Other Trade	LN-LG3	Land-large-OthPak		
C-RAIL	Rail	LN-MD1	Land-irrigated-med Sindh	H-AGW3	Waged rural landless farmers OthPak
C-ROAD	Road Transport	LN-MD2	Land-irrigated-med Punjab		
C-TRWAT	Water Transport	LN-MD3	Land-irrigated-med OthPak	H-NFNP	Rural non-farm non-poor
C-TRAIR	Air Transport	LN-SM1	Land-irrigated-sm Sindh	H-NFP	Rural non-farm poor
C-TROTH	Other Transport	LN-SM2	Land-irrigated-sm Punjab	H-URNP	Urban non-poor
C-HSNG	Rented Housing	LN-SM3	Land-irrigated-sm OthPak	H-URPR	Urban poor
C-OWNH	Own Housing	LN-DR1	Land non-irrig-sm/m Sindh	ENT	Enterprises
C-BSERV	Business Services	LN-DR2	Land non-irrig-sm/m Punjab	INSTAX	Tax to institution
C-ESERV	Education			IMPTAX	Import Tariffs
C-HSERV	Health care	LN-DR3	Land non-irrig-sm/m OthPak	COMTAX	Tax to commodity
C-PERSV	Personal Services	WATER	Water	GOV	Government
C-OSERV	Other Private Services	K-LVST	Capital livestock	ROW	Rest of World
C-PUBS	Public Services	K-AGR	Capital other agric	S-I	Capital
LA-AGL	Labor-agric (own)-large	KFORM	Capital formal	TOTAL	Total
		KINF	Capital informal		

Table C.2 Structure of the 2008 Pakistan SAM

2008 Pakistan SAM	Activities (63)	**Agriculture (30):** Wheat irrigated Sindh, wheat non-irrigated Sindh, rice-IRRI (irrigated) Sindh, rice-basmati (irrigated) Sindh, cotton (irrigated) Sindh, sugarcane (irrigated) Sindh, other field crops Sindh, fruits/vegetables Sindh, forestry Sindh, wheat irrigated Punjab, wheat non-irrigated Punjab, rice IRRI (irrigated) Punjab, rice basmati (irrigated) Punjab, cotton (irrigated) Punjab, sugar cane (irrigated) Punjab, other field crops Punjab, fruits/vegetables Punjab, forestry Punjab, wheat irrigated other Pakistan, wheat non-irrigated other Pakistan, rice IRRI (irrigated) other Pakistan, rice basmati (irrigated) other Pakistan, cotton (irrigated) other Pakistan, sugar cane (irrigated) other Pakistan, other field crops other Pakistan, fruits/vegetables other Pakistan, forestry other Pakistan, livestock (cattle, milk), livestock (poultry), fishing
		Industry (19): Mining, vegetable oils, wheat milling, rice milling (IRRI), rice milling (Basmati), sugar, other food, cotton gin (lint), cotton spin (yarn), cotton weaving (cloth), knitwear, garments, other textiles, wood and leather, chemicals, cement/bricks, petroleum refining, other manufacturing, energy
		Services (14): Trade-wholesale, trade-retail, trade-other, transport-land, transport-water, transport-air, transport-other, construction and housing, business services, education, health care, personal services, other private services, public services
	Commodities (48)	Same as activities except for the following: commodities are aggregated nationally; wheat irrigated and wheat non-irrigated activities aggregated as one commodity (wheat); wood, leather, transport-rail, transport-road, rented housing and own housing are all distinct commodities
	Factors (27)	**Labor (10):** Own-farm (large farm, medium farm Sindh, medium farm Punjab, medium farm other Pakistan, small farm Sindh, small farm Punjab, small farm other Pakistan), agricultural wage, non-agricultural unskilled, skilled
		Land (12): Large farm (Sindh, Punjab, other Pakistan), irrigated medium farm (Sindh, Punjab, other Pakistan), irrigated small farm (Sindh, Punjab, other Pakistan), non-irrigated small farm (Sindh, Punjab, other Pakistan)
		Other factors (5): Water, capital livestock, capital other-agriculture, capital formal, capital informal
	Households (19)	**Rural (17):** Large farm (Sindh, Punjab, other Pakistan), medium farm (Sindh, Punjab, other Pakistan), small farm (Sindh, Punjab, other Pakistan), landless farmer (Sindh, Punjab, other Pakistan), rural agricultural landless (Sindh, Punjab, other Pakistan), rural non-farm non-poor, rural non-farm poor
		Urban (2): Non-poor, poor
	Other institutional accounts (4)	Enterprises, government, rest of the world, capital

Environmental Benefits Statement

The World Bank is committed to reducing its environmental footprint. In support of this commitment, the Office of the Publisher leverages electronic publishing options and print-on-demand technology, which is located in regional hubs worldwide. Together, these initiatives enable print runs to be lowered and shipping distances decreased, resulting in reduced paper consumption, chemical use, greenhouse gas emissions, and waste.

The Office of the Publisher follows the recommended standards for paper use set by the Green Press Initiative. Whenever possible, books are printed on 50% to 100% postconsumer recycled paper, and at least 50% of the fiber in our book paper is either unbleached or bleached using Totally Chlorine Free (TCF), Processed Chlorine Free (PCF), or Enhanced Elemental Chlorine Free (EECF) processes.

More information about the Bank's environmental philosophy can be found at http://crinfo.worldbank.org/crinfo/environmental_responsibility/index.html.

green
press
INITIATIVE

www.ingramcontent.com/pod-product-compliance
Lightning Source LLC
Chambersburg PA
CBHW080613270326
41928CB00016B/3032